Archaeology
and the
Origins of Philosophy

SUNY SERIES IN ANCIENT GREEK PHILOSOPHY

_____

Anthony Preus, editor

ARCHAEOLOGY

AND THE

ORIGINS OF PHILOSOPHY

ROBERT HAHN

STATE UNIVERSITY OF NEW YORK PRESS

Published by
STATE UNIVERSITY OF NEW YORK PRESS, ALBANY

© 2010 State University of New York

For information, contact State University of New York Press, Albany, NY
www.sunypress.edu

Production and book design, Laurie Searl
Marketing, Michael Campochiaro

Library of Congress Cataloging-in-Publication Data

Hahn, Robert, 1952–
    Archaeology and the origins of philosophy / Robert Hahn.
        p. cm. — (SUNY series in ancient greek philosophy)
    Includes bibliographical references and index.
    ISBN 978-1-4384-3164-2 (paperback : alk. paper)
    ISBN 978-1-4384-3165-9 (hardcover : alk. paper) 1. Anaximander. 2. Philosophy, Ancient.
3. Cosmology. 4. Archaeology. I. Title.
    B208.Z7H35 2010
    182—dc22
                                                                            2009034854

              10   9   8   7   6   5   4   3   2   1

For Amy, Zoë, Chava
who graciously allowed me to go to Greece
year after year, for weeks at a time,
to prepare this book

Contents

PART II.
ARCHAEOLOGY AND THE METAPHYSICAL
FOUNDATIONS OF AN HISTORICAL NARRATIVE
ABOUT THE ORIGINS OF PHILOSOPHY

Illustrations

Preface

In May 2000, I had the honor of giving the Theophilos Beikos Lecture at the University of Athens. My presentation was a summary of a central theme of *Anaximander and the Architects* that was then just about to be published. By chance, one of the most senior scholars in the field, Alexander Mourelatos, was in the audience and after the lecture he made a remark to me that, unforeseeable at that time, led to this new book. Up until that time, I had been describing my research as trying to show how an appeal to ancient architecture and its technologies could illuminate Anaximander's thought and a fortiori the origins of Greek philosophy. Mourelatos remarked that he had never before heard anyone explain Greek philosophy by appeal to *archaeology*. I was initially taken by surprise by the way he put it. But I realized straightaway that my research in ancient architecture and its technologies relied on archaeological artifacts and archaeological reports. And so while I was struck by his wording, I also was struck by the very idea that despite the enormous literature on Greek philosophy, scholars have not appealed to archaeology for major insights. My next book project, *Anaximander in Context* (2003), continued architectural themes but also included discussions of proportions and numbers in archaic sculpture (and the metrics of poetry), and so my appeal to archaeology was extended further.

An additional review of the secondary literature has confirmed that the approach I have taken is unique, and with that originality comes a host of difficulties. In the absence of scholarship taking this approach, I had no exemplars to follow or critique. At first, I was not sure how to articulate the methodology that I in fact employed in these earlier studies. But after much reflection I came to conclude that the method could be presented in a simple and straightforward way. In this new book, I propose and articulate that method for future research; that method has two distinct parts: a "Method of Discovery" by which, given the doxographical reports that mention technological analogies and metaphors, we must imagine an ancient thinker watching and reflecting on these very technologies, the evidence for which comes from the archaeologist, and a "Method of Exposition" by which the arguments about archaeology's relevance can be demonstrated. The Method of Exposition is threefold: (A) present the doxographical testimony to which scholars routinely appeal when explicating pre-Socratic thought, (B) assemble the debate among scholars about the textual and interpretative difficulties, and (C) appeal to archaeological techniques, reports, and artifacts to show how the older and conventional approaches to the text and its interpretation can be clarified and illuminated.

The success of this volume, then, rests on whether we can answer affirmatively two questions: (a) Do the appeals to archaeology illuminate the technical analogies and metaphors mentioned in the doxographical reports, and, in so doing, illuminate the culture in which these early philosophers lived and thought and consequently the historical context in which their ideas achieve a new range of meanings? And (b) Do we have more *philosophical* insight into the texts and interpretative difficulties by the appeal to archaeological artifacts and reports?

The main body of pre-Socratic studies reads something like a debate over conversations that emanate from "brains-in-jars." By this I mean that the scholarly debates proceed as if the pre-Socratic thinkers did not have *lives*, that their thoughts were not fashioned in a complex process of integrating experiences in their youth, the goings-on and contingencies in their communities as their lives flourished, along with intellectual discussions that were generated and anchored in a comparable manner. This is, of course, a consequence of the fact that the evidence about them is exiguous, and the surviving biographical summaries are very distant from the time of their flourishing, often fantastic and hard to rely upon. But, of course, they did have lives, and one small window I hope to open by this book is the one that looks on hands-on activities displayed in front of their very eyes, in their historical and cultural contexts, alluded to by their technical analogies and metaphors. The archaeologists offer to supply some of these details through this opened window. I do not mean to suggest that the generations of scholarly debates have a diminished value, but rather that appeals to archaeological reports and artifacts offer another resource for insight, contribute to these familiar approaches, and have been routinely and systematically overlooked. The objective of Part II of this book is to explore just why it is that scholars have not pursued archaeology as a resource for ancient philosophy. The simple answer, as the reader will discover, is that the conventional approach to ancient Greek philosophy has proceeded by way of a myopic view of what counts as "philosophically appropriate." Scholars have missed this resource because they have entered into the field with conceptual "blinders," obfuscating the importance and relevance of historical and cultural factors. Let us take off these blinders and allow new rays of light to shine in.

This book has been a project taking shape over the last eight years and, consequently, there are many people to acknowledge who were helpful along the way. This is not to say that everyone I consulted or with whom I discussed various parts of the project necessarily agreed with my approach or the details. And it is also noteworthy, as it will seem obvious to most authors, that sometimes even a few casual comments proved to be most influential in the development of lines of thought. First, I would like to acknowledge the many people, young and old, who enrolled in my annual travel seminars to Greece, Turkey, and Egypt, over almost thirty years, who cheerfully listened to what I had been thinking and writing about, and engaged me in useful and memorable ways. I wish to recognize with special gratitude my graphic artist, Cynthia Graeff, who has assisted me with the illustrations in my last three book projects, in addition to collaborating with me for almost fifteen years, designing visual materials for my Greece and Egypt projects. Since so many of my arguments are advanced by visual evidence, I am deeply indebted to and dependent upon her fine work. I wish also to thank Theresa Huntsman, who helped me organize and secure the many requests for permissions to reproduce copyrighted materials and to review parts of the manuscript. I would like to acknowledge gratefully the help in preparing an index from my graduate assistant Tim McCune. And I would like to acknowledge the support and kindness of my department chairperson, George Schedler, and Alan Vaux, dean of the College of Liberal Arts at Southern Illinois University, who, along with the administration, approved sabbatical leave to organize my research and finish this book.

I would like to acknowledge archaeologists who greatly affected my work. First, Manolis Korres' work deeply influenced the way I have thought about technologies at the ancient building sites; Korres graciously allowed me to reproduce some of his extraordinary drawings in this book. Kathleen Lynch helped me more than I can ever adequately thank her for, traveling with me in Greece, Turkey, and Egypt, preparing research reports that I have drawn on freely, and enabling me to understand how art historians and archaeologists argue their cases. I also owe a debt to Barbara Tsakirgis, who traveled with me in Egypt and replied many times in most useful ways to my inquiries about archaic Greece. I owe much to Peter Schneider, who drew my attention to details in Ionian building techniques in Didyma, Miletos, and Ephesos, and who discussed my ideas at length since the mid-1990s. Anton Bammer and Ulrike Muss escorted me through the Artemision in Ephesos several times and in most illuminating ways, and met with me in Turkey over the years to listen to and respond to my ideas; I am most grateful to them. Nils Hellner was kind enough to allow me to use his wonderful drawings depicting techniques at the Samian Heraion, showed me significant and relevant ancient evidence, and often clarified important details that I had not understood fully. I am most appreciative for the exchanges I have had with Mark Wilson Jones, whose work about the origins of the Greek orders inspired me to reflect again on the kind of narrative I was proposing for the origins of philosophy, and whose comments along the way were very helpful. I would like to acknowledge gratefully Prudence Rice who read my manuscript and made many helpful suggestions. And I would like to record my sincere thanks to Hermann Kienast, who, over the course of more than twenty-five years, continually objected to almost everything I proposed—I am still not sure that he has understood my argument that Anaximander came to imagine the cosmos as cosmic architecture. Hermann always insists that he is a craftsman, not a philosopher. Nevertheless, Hermann was always willing to discuss the details of ancient architecture and building techniques, gracious enough to introduce me to Gottfried Gruben, who made a few very useful suggestions, and patient enough to lead me through the Samian Heraion and the Eupalinion so that I might gain more insight into some of my many questions that remained unanswered, even after I had translated large portions of his deservedly famous studies.

I also am indebted to many philosophers and classicists. Classicists Rick Williams and David Johnson traveled with me to Greece on occasion, corrected many of my more egregious philological errors, and kindly read my work with care. My Greek colleagues in Athens—Leonidas Bargeliotes, Yiorgos Maniatis, and Soteres Founarou—graciously discussed parts of my work when I met them in Greece many times. And I owe an extraordinary debt over the last decade to classicist Nancy Ruff, my co-director of the Ancient Legacies seminars, who patiently listened to my developing thoughts while we traveled in Greece, Turkey, and Egypt, and who read through my last three manuscripts. I am grateful to philosophers Randy Auxier, Douglas Berger, David Clark Jr., and Jason Rickman who led me into a world of approaches to historical narratives in a way that I had not quite visited before, reading some of my work and commenting on my efforts to grasp the insights of Collingwood, Dewey, Peirce, and James. Gerard Naddaf and Dirk Couprie continued to inspire me by their work on Anaximander, and Greek philosophy in general, and, though their approaches are quite different, both have helped me to think more deeply about Anaximander over the last fifteen years; it is not easy to thank them sufficiently. John Kaag aided me enormously with the Herculean task of interweaving my archaeological research with the philosophical traditions of twentieth-century hermeneutical and pragmatic approaches; the structure of the argument in chapters nine and ten bears his imprimatur. And finally, it is an honor and a pleasure to acknowledge Mark Johnson, who has assisted me in manners great and small over the last quarter-century, who traveled with me in Greece and Egypt at the beginning of my projects and again at the end, and who helped me understand what I was doing philosophically at times when I was confused by my own research. Johnson's work on the bodily basis of meaning continues to exercise great influence on the way I think about my work, and helped me to express more clearly yet the kind of "experiential realism" to which my work is committed.

In a class that requires to be set up all by itself, I wish to record my gratitude and my appreciation to Tony Preus, who took an interest in my research as I first fumbled with it in the late 1980s, and whose continual encouragement and support saw me through to my third book in his series.

Acknowledgments

Grateful acknowledgment to: Manolis Korres for permission to print illustrations Scaffolding around the Parthenon (fig. 1.6), Workmen at the quarry (fig. 1.7), Removing the blocks from the quarry (fig. 1.8), Transporting blocks from the quarry to building site (fig. 1.9), Transporting stones up an incline by means of a team of oxen that pulled a rope drawn cart, circling a heavy post (fig. 1.10), Lifting devices for installing column drums (fig. 1.11), Reconstruction of the great crane used to complete the installation of the upper orders of the temple (fig. 1.13), and The stages of geometrical imagination preparing a column capital, using straight-edge and caliper (fig. 1.14); Christina Kolb for permission to print illustration Reconstruction of Hephaistos with bellows by Christina Kolb (fig. 1.3 top); Aenee Ohnesorg for permission to print An unfinished column drum with boss, from the archaic temple to Artemis in Ephesos, dating to the mid-6th century BCE (fig. 1.12); Staatliche Museum, Kassel for permission to print Lakonian Hunt Painter cup (c. 565-550 BCE), Sisyphus in the underworld, with a column (fig. 2.1); Gay H. Cone for permission to print Kleophrades Painter (c. 500 BCE), Sisyphus with Hermes and Cerberus, with a column (fig. 2.2); Art Resource, New York for permission to print Lakonian Arkesilas Painter (c. 550), Atlas and Prometheus supported by the upper part of a column (fig. 2.3), Berlin Foundry Cup depicting end stages in the production of a bronze figure, side A (fig. 4.6), Berlin Foundry Cup depicting end states in the production of a bronze figure, Side B (fig. 4.7), Berlin Foundry Cup depicting end stages in the production of a bronze figure. Blacksmith wears skullcap and workman, with rare frontal face, works the bellows (fig. 4.8), Gigantomachy from the Siphnian Treasury in Delphi (figs. 1.3 center, 4.3 top); American Journal of Archaeology for permission to print Gigantomachy from the Siphnian Treasury in Delphi ca. 525 BCE, Reconstruction drawing of scene, after Moore (figs. 1.3 bottom, 4.3 bottom), Preindustrial smelting technique (fig. 4.1 a, b, c), Reconstruction of a double pot bellows made of clay and covered with skin (fig. 4.12); the Acropolis Museum, Athens for permission to print Black figure kantharos ca. 550 BCE, depicting Hephaistos with the double bellows (fig. 4.2); the Bibliothèque Nationale de France for permission to print Red figure cup by Douris ca. 475 BCE (fig. 4.4 a, b); the Museo Archeologico di Caltanissetta for permission to print Column krater by the Harrow Painter (fig. 4.5); the American School of Classical Studies Athenian Agora Excavations for permission to print Nozzle of a bellows, mid-6th century BCE, from the Athenian Agora (fig. 4.9), In the Roman period Athenian blacksmiths re-used the narrow necks of amphorai as nozzles (fig. 4.10) Deutsche Archäologische Institut for permission to print illustration of Bellows nozzle preserved from an archaic bronze working shop under the later "Workshop of Pheidias" at Olympia (fig. 4.11); Philip Betancourt for permission to print Nozzle of the bellows from Chrysokamino, Crete dating to late Neolithic or early Minoan times (fig. 4.13), Reconstruction of the use of the bellows bladder from Chrysokamino, Crete dating to late Neolithic or early Minoan times (fig. 4.14); the Israel Museum for permission to print Qumran Sundial (fig. 6.5); Nils Hellner for permission to print illustrations Trisecting a column drum face (i.e. a circle) by Nils Hellner(fig. 6.16), Fragment of a column drum from the archaic temple to Hera in Samos, Dipteros II, exhibiting markings for trisecting the drum face by Nils Hellner (fig. 6.18). And I wish to make a special acknowledgement to Dirk Couprie, who brought to my attention following publication an important error in my illustration of the plan view of Anaximander's cosmos, Figure 1.2, repeated again as Figure 3.5. In addition, I am also grateful to him for a small correction in the plan view reconstruction that I propose for Anaximander's sundial, Figure 6.11.

Introduction

I wrote *Archaeology and the Origins of Philosophy* in light of my earlier books but with the intention of not requiring my readers to have familiarity with them. My ambition here is to explain the methodologies I have employed in two earlier works, *Anaximander and the Architects* (*A&A*, 2001) and (co-authored) *Anaximander in Context* (*AiC*, 2003), with the express purpose of offering as a paradigm a new model for future research in ancient philosophy. Part I of this book presents case studies drawn from, but also significantly extending, my earlier work. Part II of this book engages in an articulation of and reflection about the methodologies used in these case studies and the paradigm they offer to add to the study of ancient philosophy. I continue to embrace the thesis that Anaximander was present at the temple building sites and witnessed the technologies there that he applied imaginatively to his cosmic speculations. And thus to the readers already familiar with my earlier work who might start with Part I and think "Haven't I already read something like this?" I recommend that they read Part II *first* and only then turn to examine the details of the case studies presented in Part I. The general thesis about Anaximander and architectural technologies is again central to this study, but the case studies are new and attempt to provide exemplars organized in accordance with these new methodologies.

Let me also make clear that it is not my intention to diminish the long successful approaches in ancient philosophy but rather to offer alongside the familiar styles of scholarship this new research model. The extensive reflections on the metaphysical foundations of these historically and culturally embedded case studies (in Part II) seemed appropriate only in the light of these case studies (in Part I) showing that archaeological resources really are capable of illuminating the abstract and speculative thought that is central to the scholarship on ancient philosophy.

PART II—ARCHAEOLOGY AND THE
ORIGINS OF PHILOSOPHY: AN ARGUMENT
FOR THE PHILOSOPHICAL IMPORTANCE OF
HISTORICAL AND CULTURAL CONTEXT

Chapter seven sets out the problem and initially responds to the question: Why have studies of ancient Greek philosophy and its origins routinely neglected archaeological resources? The argument revolves around Jonathan Barnes, who despite all the marvelous studies he has produced, presents a myopic vision of what is philosophically appropriate. Barnes' two-volume work on the pre-Socratics is arguably the most influential work in the field for the last fifty years. By his vision of what counts as philosophy, it is no wonder why appeals to archaeological evidence are routinely neglected. In his first edition, Barnes insisted that "philosophy lives a supracelestial life beyond the confines of space and time," and when he was criticized for his historical insensitivities, he reinforced his earlier view, in his second edition, to rule out of court studies that urged consideration of historical and cultural context. Barnes' work has been so influential; his view is only the most extreme among so many studies that share his inclinations about what counts as "philosophical." But this point of view undermines the very appropriateness of research that appeals to archaeology, research that must take seriously historical and cultural contexts. As such, if the case studies I have presented here proved to be philosophically illuminating, it is fair to conclude that Barnes' approach is too extreme, and this means (what seems so obvious to me) that historical and culture context are also significant to research in our field.

Chapter eight traces out the development of archaeological approaches over the course of the last century. The underlying foci are to explore the theoretical frames in terms of which the archaeologists offer their narratives, and to try to understand *how* archaeologists have attempted to infer abstract thoughts from artifacts—how historical and cultural context assumes a place in understanding mentalities. In a curious way, as I see it, both Anaximander *and* I are standing in front of the same "architectural" evidence and reflecting upon matters that have now become the business of the archaeologist. So, what can we learn from the archaeologist? How does the archaeologist infer abstract ideas from the material culture? Accordingly, the survey of old archaeology, new archaeology, processual, postprocessual, and cognitive archaeology is designed to show to what extent abstract and speculative thoughts are implied by material artifacts, and how postprocessual archaeologists who champion "interpretative archaeology" undermine the meaningful distinction between "evidence" and "interpretation."

Chapter nine explores the "imaginative meaning of an artifact," and traces out the ideas of hermeneutic play and interpretation as applied to archaeology. I have been arguing that Anaximander came to imagine the cosmos by means of architectural techniques; what are the imaginative dimensions of determining the meaning of artifacts, indeed, objects in general? This discussion begins with Gadamer's work, and then turns to pragmatic interpretations of Dewey, James, and Peirce to explore a philosophical defense of how we infer from material context to abstract, imagined thought. All these positions, taken separately and together, stand in marked contrast to Barnes' positivism; they all insist that artifacts are never just artifacts, that the objects of the world cannot be understood by way of a single objective meaning but only through a process of imaginative interpretation. This theme is then connected to Quine, whose work on the indeterminacy of translation shares this commitment, and then to Davidson's radical interpretation that offers too extreme a position that can be countenanced no more than Barnes' positivism. Next, Putnam's internal or historical realism is considered because his position also offers a way to make sense of claims being true, and some opinions being "better" than others, without succumbing to either positivism on the one hand or contextual relativism on the other. Putnam's arguments offer a way to show how we can place constraints on a narrative without embracing the correspondence theory of truth; Putnam places truth claims within an historical context, but not a "trans-temporal context" as Barnes suggests. And finally, those positions are contrasted with Searle's arguments for the existence of brute facts, as opposed to institutional facts, as he argues for a version of metaphysical realism that he calls external realism. I argue against Searle and metaphysical realism in general.

Chapter ten begins by setting out an archaeological approach to ancient thought, discusses James and Dewey on the context of consciousness to set the platform for the natural and material basis of mental life, and then turns to a presentation of metaphor and bodily experience following the work of Mark Johnson and George Lakoff. To place Anaximander at the building site as a careful observer of temple architecture, who imaginatively projects the techniques he witnesses onto the cosmos because he envisions cosmic architecture, is to place bodily experience at the center of abstract and speculative knowledge. And to place bodily experience at the center is to argue for the importance of historical and cultural context for our understanding of (some aspects of) ancient philosophy and against Barnes' extreme view, the supracelestial vision of philosophy.

Chapter eleven explicitly sets out the new methods I have employed. My research has two parts: what I am calling a "Method of Discovery" by which we recreate in a meaningful way the historically and culturally embedded experience of Anaximander, or some other ancient thinker, and a "Method of Exposition" by which we promulgate the arguments that make our case.

The key to the Method of Discovery is to isolate in the doxographical reports the references to ancient technologies and techniques, then to appeal to archaeological reports and artifacts to recreate the ancient processes and products delineated by archaeologists, and finally to connect the doxographies to these archaeological reports to illuminate a range of experience that Anaximander and other ancient thinkers would have plausibly experienced. What this method amounts to, in my estimation, is that (in Anaximander's case) when we recreate the archaeological reports on temple building, we are standing next to both the archaeologist and Anaximander (i.e., the Anaximander whose belief system we know through the doxographical reports) in the presence of the surviving archaic artifacts. It is only by recreating the activities going on at the building site that we can come to grasp the illuminating historical and cultural context of the doxographical reports. This approach is what I learned from Collingwood in his *The Idea of History*.

The key to the Method of Exposition is nothing other than clarity of presentation. This method has three parts. In the first part (A), I set out the relevant doxographical reports, the locus of the evidence, and then (B), I present the scholarly debate about this evidence. Finally, (C), I appeal to archaeological resources to clarify or resolve the old debates.

I believe these methods can be employed to facilitate new research in the field. I hope to follow this method in future studies on the pre-Socratics, Plato, and Aristotle. Consider the possibilities that every reference to an ancient technology or technique—throughout our ancient philosophical literature—might find a report showing evidence of the ancient artifacts with detailed explanation about the processes by which the artifacts were produced. We will find, I believe, that, in many cases, the appeals to this kind of historical and cultural context would be useful and revealing.

Chapter twelve weaves together archaeological developments and philosophical approaches. In summary form, I make clear that I am championing a realism that dismisses the distinction between "evidence" and "interpretation." A fact is a posit from which certain things follow or to which they are connected; they are not chiseled in stone, nor can they claim meaningfully to be "true" in a transhistorical or supracelestial sense. We must understand that an historical narrative requires the selection of a starting point and an ending point, and proposes a causal account that offers to connect the "facts." By the purposes and values that direct specific inquiries, our narratives are produced. Consequently, they are not copies or pictures of some antecedent reality. In just this sense, the age-old problems set by the program in metaphysical realism are rejected as hopeless and wrongheaded; an appeal is made to Kant's *Critique of Pure Reason* as an historical transition point in rejecting that program and the correspondence theory of truth that is its cornerstone. Finally, I offer a brief reflection on the reality of the past, contrasting it with Barnes' supracelestial thesis; the past has a meaning in the creative present. The project of reifying the past, and trying to make our narratives map onto that antecedent reality—the correspondence theory of truth—is, consequently, seen to be metaphysically mistaken.

PART I—ANAXIMANDER'S COSMIC PICTURE:
THE CASE STUDIES

Chapter one has three parts. The first part is an historical narrative that exhibits the results of what I am calling the new "Method of Discovery"; the narrative does not pretend to describe all of Anaximander's life, or cover all of his thoughts. The narrative is a glimmer, a slice of his life in eastern Greece that unearths some of his thoughts and the processes by which they burgeoned in his mind. It places Anaximander at the temple building sites, and situates those projects in the historical and cultural context of the sixth century BCE. The second part of chapter one contains five illustrations, each of which introduces the highlight of each of the following five chapters. The "picture" is the basis of the argument; each presents an archaeological resource as a clarifying answer to a traditional problem in Anaximander scholarship. The third part of chapter one invites the reader to take an imaginative journey to an ancient temple building site and, thanks to the fabulous illustrations made by Manolis Korres, who has graciously given me permission to reproduce some of them, our philosophers will be positively amazed by evidence that they have routinely overlooked. The ancient building sites and their related technologies provided the Greeks of the archaic and classical periods with a veritable experimental laboratory where principles of nature were displayed, explored, and tested, and where abstract, symbolic, and imaginative thought were realized in concrete forms.

Chapter two explores the size and shape of Anaximander's earth. In A&A, the investigation focused on column-drum preparation in the techniques of *anathyrôsis* and *empolion*. In AiC, the investigation concentrated on technical issues of where to measure "(lower) column diameter," that is, the architect's building module. In this new chapter, the art historical record of vase painting is explored (something I had never done before) to show the evidence that the column and column drum already had symbolic meaning in the archaic period. Then, the debate is extended, in light of the archaeological evidence, to determine where on the column was the 3×1 drum that Anaximander imagined analogous with the shape and size of the earth. The module is usually identified with "lower column diameter," but there is no "metrological rule" for determining drum proportions. This means that drums could be of *any* proportions (i.e., 2:1, 3:1, 4:1, 5:1) so long as, when stacked together, they finally reached the same height for the installation of the capital. However, the archaeological evidence also shows that there were, in fact, column *bases* in the Ionic temples that were metrologically determined (i.e., they were all exactly the same proportions), and we have evidence for 3×1 exemplars. Did Anaximander imagine the column-drum earth as a column base? After reviewing the fascinating archaeological evidence, I argue that it makes more sense to suppose that he did not, but that was because he imagined the earth

aloft, held up by nothing. The column base purposefully obfuscates this crucial point since it rested on the earth-foundation.

Chapter three explores two other features of Anaximander's cosmic
picture: the earth in equilibrium and the cosmic numbers. In A&A, the
issue of the earth's equilibrium was clarified by appeal to the architect's
techniques of plan and elevation views; in AiC, the cosmic distances to the
stars, moon, and sun were clarified as increments of the archaic formula of
9 and 9 + 1 embraced by Homer, Hesiod, and others. In this new chapter,
the distinctions between plan and elevation views are again emphasized,
but this time the distinction between interaxial versus intercolumnar measurements plays the pivotal role. And the defense of the poetic formulas of
9 and 9 + 1 is now reinforced by a long discussion of how the archaeologist
understands "Technological Style" and "Technological Choice," which are
examples of historical and cultural context as important factors in the work
of archaeologists.

Chapter four examines another aspect of Anaximander's cosmic picture
and shows that *prêstêros aulos*, the nozzle of the bellows, is the correct reading of how the fire radiates from the cosmic wheels, because, in this hylozoistic universe, the cosmos is alive by breathing. Neither A&A nor AiC has
any detailed discussion of this complex issue. In this new work, however,
there is a long exploration of the archaeological evidence for the bellows.
The central theme of the new illustrations includes Hephaistos carrying
around his bellows, which is an animal skin—sometimes a double animal skin
that suggests a pair of lungs—to be inflated and deflated like a breathing animal. The result of this study is to make the case that the bellows displayed a
breathing mechanism; Anaximander projected imaginatively on to the living
cosmos an instrument of breathing from the metal workshops at the building
site. It turns out that the cosmos is alive as an everlasting fire-breather, a doctrine that is usually attributed to Heraclitus.

Chapter five investigates another aspect of Anaximander's cosmic picture, how the stars, moon, and sun are really heavenly wheels of fire encased
in mist. Had Anaximander seen any vehicles of transport with hollow-
rimmed wheels? In AiC, there is no discussion about wheels, neither cosmic
nor land vehicles. In A&A, there is passing reference to the architect Metagenes' wheeled vehicle for transporting monolithic architraves from quarry to
building site. What the new chapter argues at length (and not in passing) is
the most likely candidate for Anaximander's cosmic imagination. In this new
work, Metagenes' invention is reached finally at the end of a long discussion
of the evidence for ancient wheeled vehicles. These techniques of making
wheeled vehicles, curiously enough, have implications for both Anaximander's sundial and his map of the earth. And when this discussion of Anaximander's wheels is placed in the wheelwright's workshop, the wheel-and-axle
construction suggests unmistakably a cosmic axle, an *axis mundi*. To unfold

this imagery, the discussion of Anaximander's cosmic wheels is placed in the context of Pherecydes' account of the cosmos as a great tree.

Chapter six attempts to reconstruct Anaximander's seasonal sundial. The sundial is not considered at all in *AiC*, and in *A&A*, while a reconstruction is proposed, the model is expressly rejected at the opening of this chapter. The new, proposed reconstruction of the sundial is shown to mediate between the cosmic picture (or cosmic map) and the map of the earth. The earliest surviving sundials in Greece date to Hellenistic times when conical and hemispherical sundials make sense, reflecting spherical conceptions of the earth's shape. But what did Anaximander's sundial look like, given his conception of a column-drum–shaped earth? In the new reconstruction, a column drum is proposed as the dial face with a gnomon placed vertically, and implications for his map are also considered in terms of the shadows cast. After the details have been considered, important objections are raised and answered. Contrary to the pronouncement of Cornford and others, there are good reasons to suppose that "observations" played a significant role in some (but not all) of Anaximander's cosmological speculations.

Thus, *Archaeology and the Origins of Philosophy* proposes a new additional methodology for research in ancient philosophy; as a volume intended to "stand alone" (and not presupposing familiarity with my earlier publications), Part II, the philosophical argument for the relevance of archaeology and the importance of historical and cultural context, needs Part I, the series of case studies in the newly articulated methodology. If the readers can see that we really do learn new things, and rich details, from archaeological resources that shed new light on Anaximander's abstract and speculative thoughts and his thought processes (and, in future projects, other pre-Socratics, Plato, and Aristotle), that is, Part I, then it is serious and worthwhile to explore how and why this is so, and why it should have been missed.

I had been trying for a long time to figure out why my interpretation had been so new, and why not just this theme but the whole approach of ancient philosophy routinely and systematically ruled out appeals to archaeology. In a curious way, I was lucky that so prominent and distinguished a scholar as Jonathan Barnes could help me get clear about what has gone wrong. My approach to Anaximander has been to place him in his cultural context; the kind of approach championed by Barnes and those who share his "trans-temporal" view undermines the appropriateness of this kind of study. If we can see, after reading this book, that we do gain insights into Anaximander's abstract and speculative thought, and, furthermore, that these insights are philosophically relevant, we should realize that traditional, and perhaps unspoken, guidelines limiting the scope of philosophical research need to be revised.

PART I

Archaeology and Anaximander's Cosmic Picture: An Historical Narrative

ONE

Anaximander, Architectural Historian of the Cosmos

A

WHY DID ANAXIMANDER WRITE A PROSE BOOK RATIONALIZING THE COSMOS?

WE KNOW THAT Anaximander wrote at least one book, a *prose* treatise, since a fragment of it has survived, preserved by Simplicius.[1] Since the writing of Cornford's *Principium Sapientiae*, we have an adequate identification of Anaximander's book: it rightfully attains a place not only as the first philosophical book in prose but also as a *rationalized* version of Hesiod's *Theogony*, an Hellenicized version of the Babylonian creation story, the *Enuma elish*.[2] These earlier poetical narratives tell of the origins of the cosmos and its series of developments to the present world order. The surviving doxographical fragments show that Anaximander's book explored these same matters.[3] His book plausibly began with the origins of the cosmos and included the stages through which the cosmos developed, probably to his present day.[4] Like the earlier works, as Cornford established and Vernant echoed,[5] Anaximander too supposes an original and primordial unity that is the common origin of all things and from which things separate out; there is perpetual motion in the cosmos; and there is a constant struggle against and uniting of oppositions in the stages of cosmic development. In all three cases, the thought is cosmogonical, that is, the state of things now is not the way it was at the very beginning, and so an account is supplied to explain how things got this way, that is, the stages of development that connect the origins with the present. But Anaximander's narrative is in *prose*, not in the poetic form of the *Theogony* and *Enuma elish*, and it is a *rationalized* account, not a mythological one, of those origins and developments. Absent from

Anaximander's account are the familiar references to gods and goddesses; instead we have the articulation of cosmic structure—the stages by which the cosmos was built, the stages by which earth, sun, moon, and stars were formed sequentially—and in a way unappreciated in much of the secondary literature, of cosmic mechanism.[6] Two questions that remain largely unanswered by classical scholarship are: (1) Why did Anaximander write in prose? (2) Why did he *rationalize* the cosmos instead of mythologizing it as earlier cosmogonists had done? Asked differently, what convergence of factors led him to his natural, not supernatural, narrative? If we are to supply an adequate answer to these questions, granting the exiguous nature of the doxographical evidence, we may look profitably to the historical and cultural context in which Anaximander lived and flourished.

The well-known doxographies of pre-Socratic thinkers, compiled masterfully by Hermann Diels and worked over by generations of scholars, neglect to provide the historical and cultural contexts in which and by means of which these "opinions" emerged. However, it is possible to reconstruct that context for Anaximander, and what this means for us is to think through for ourselves, and so imaginatively interconnect, the background events to the surviving fragments, and so to illuminate them. These "opinions" do not exist in a vacuum, of course, but were part of a life, and it is in the context of this life that the thoughts acquire a meaning, not only for Anaximander but also for his Ionian compatriots who were his audience. In the case of Anaximander, the missing background included a series of extraordinary architectural events that overwhelmed his archaic community and informed and illuminated his cosmic thought. In this context, it becomes clear that Anaximander's prose account conveys the development of the cosmos in architectural stages. The case here is not that Anaximander was an architect but rather, watching the architects work, Anaximander was inspired to become an *architectural historian of the cosmos*. Commentators have long regarded his book to be a kind of *historia* of the cosmos, from the very beginning to his present day, along the lines of the Babylonian *Enuma elish* and Hesiod's *Theogony*, but what they missed was the architectural approach that shaped his speculative and rationalizing thoughts on the matter. The missing historical and cultural background events were the ones that unfolded in front of his eyes at the great building sites where temples of hard stone and of gigantic proportions were produced for the first time in Ionia in the sixth century BCE.

When we know what happened at the temple building sites—what the architects did and how they did it—we will have before us the clues for reaching compelling hypotheses[7] about why Anaximander wrote a prose treatise and why he expounded a rational account of the cosmos. At the building sites, the Ionian architects built cosmic houses and wrote prose books about them.[8] They planned and executed *rationally* the multitasking enterprises required to produce successfully these gigantic stone temples, and they communicated and celebrated their achievements in prose books by recording their rational strategies in honor of cosmic themes. The architects could not adequately convey in the language of poetry as they could in prose the rules of proportion for temple building or the details involved in the construction process itself. Anaximander's rationalizing prose book—which can be dated to 548 BCE[9]—responded to and acquired a meaning within and against this context.

The archaic architects of the great Ionic temples—the only names that reach us are Theodoros (and Rhoikos) of the temple of Hera in Samos, and Chersiphron and Metagenes of the temple of Artemis in Ephesos—are the only other authors credited both contemporaneously and in the same region, with the single exception of Pherecydes of Syros, with writing the earliest Greek prose treatises. Vitruvius tells us that Theodoros published a book on the Samian Heraion, and that Chersiphron and Metagenes published one on the Ephesian Artemision. And we know that Theodoros's temple in Samos was begun circa 575 and the temple of Chersiphron and Metagenes was certainly under way circa 560,[10] in any event before the publication of Anaximander's book in 548.[11] From reports that Vitruvius presents about the rules of proportions for the temples, and technical innovations they achieved—especially the wheeled devices that Chersiphron and Metagenes invented to move the large blocks and architraves—we have good reasons to suppose that these themes were discussed in their books.[12] When Xenophon in his *Memoribilia* makes Socrates ask the sophist Euthydemus, who has made a collection of books, whether he is studying to be an architect, the sarcasm would be pointless unless it was common knowledge that architects wrote books and were accustomed to learn by means of them.[13] That tradition is believably traced to these architects of archaic Greece. Given this evidence, we know of only two early sources of prose writing produced in *eastern* Greece dating to the mid-sixth century BCE: Anaximander and the architects.[14]

The building programs of monumental temples transfixed the archaic communities of Ionia in a manner unparalleled by anything other than war and the preparations for it. The great temples quite literally transformed their very landscape and the horizons in terms of which they grasped their visions and capacities. Anaximander was responding to and within the community of activities inaugurated by the architects; he took his own clues from the architect's temple building site: in short, he came to imagine the cosmos as a kind of cosmic architecture. And, moreover, we should see in this *agôn* culture, a culture that celebrated human excellence and claims to it, that Anaximander did the architects one better; while the architects wrote in prose about building the temple, the house of the cosmic power, Anaximander wrote prosaically about the house that *is* the cosmos, that is, as architectural historian of the cosmos.

Behind the scenes of building these gigantic stone temples, we must imagine Anaximander and his community discussing and debating the house appropriate to represent the cosmic powers—Apollo in Didyma, but also Artemis in Ephesos, and Hera in Samos. After all, the temple was the house of the god. What architectural structures would appropriately represent these abstract cosmic powers? What symbolic numbers, proportions, and designs would idealize the highest and best? And then the architects had to modify the communal "wish list" by reconciling symbolic design with the structural demands that nature imposed. Of course, certain materials had tensile limits; for instance, marble can span greater lengths than limestone while still carrying a greater load on the upper orders of the temple. Many such "truths of nature" that the architects' work exhibited constrained the imaginative vision of the community. Under the leadership of the patrons, and the architects who worked at their pleasure,[15] decisions had to be made, then, as to how best to represent symbolically these cosmic powers, given nature's constraints. In the process, Anaximander saw how the abstract powers were made concrete in the sequences of hard stone structures that gave expression to them. Anaximander projected *imaginatively* these ongoing discussions, debates, and reflections about how to visualize and concretize abstract, cosmic powers; he imagined the cosmos itself as built-architecture and sought to explain, on analogy with the architects, the stages by which the cosmos itself was built. Thus, when we attend to his historical and cultural context, we know why Anaximander wrote a prose book and rationalized his narrative. In the context of the architects' building and prose writing, Anaximander transformed the older poetic mythologies of the *Theogony* and *Enuma elish*. Anaximander emerged as a writer of a cosmic architectural narrative; the prose and rationalizing character of his writing joined company with, and perhaps rivalry to, the writings of the architects.

Anaximander identified the shape and size of the earth with a column drum whose diameter was three times its depth, that is, whose proportions were 3 × 1. Most commentators never followed up on that architectural analogy, never investigated column-drum construction, never provided a photo or illustration of a surviving archaic column drum to see if any more clues to Anaximander's thought might be found in such an investigation, but instead preferred to see the claim as a throwaway, a casual comment made merely to add vividness to Anaximander's abstract speculation that the earth's shape was flat and cylindrical. However, when we follow this architectural hint and explore the temple building sites where we will find in fact 3 × 1 column drums, a rich story unfolds. We discover, first of all, that column-drum construction out of hard stone was new to Ionia at just the time Anaximander lived and flourished, and was instrumental, indeed indispensable, to the construction of the great monumental temples to

Apollo in Didyma, Artemis in Ephesos, and Hera in Samos. Without this innovation of column-drum construction in archaic Ionia, monumental temple building would not have been feasible, since it was impossible to deliver safely from quarries often many kilometers away the hundreds of monolithic columns required, each stretching lengths of more than fifty feet. Perhaps, then, Anaximander's identification of the earth with the column drum was no throwaway at all, as scholars must have assumed by their silence, but rather the so-called tip of the iceberg, a hint of or clue to the things going on at the temple building site that fascinated, challenged, and inspired Anaximander and provided a background context for his abstract cosmic speculations. The success of monumental temple building required a rational, not poetic mentality; the many proportions[16] and detailed techniques described in books by the architects could be conveyed best in prose not poetry, and the stages of building set out step by step provided an exemplar of rational thinking that inspired Anaximander and his would-be audience in the archaic Ionian communities.

As we explore further the activities at the building sites, we begin to imagine the technological know-how that was required to accomplish successfully these enormous feats. We can imagine what Anaximander would have witnessed in what was quite literally his own backyard. Special techniques were required to unearth stones from their quarry bed and special tools were invented for the job; lifting devices of specific and gigantic dimensions were required to move the multi-tonned stones; vehicles of transport with reinforced wheels and substantial axles were needed to convey the heavy stones; and more lifting devices, levers, set-squares, leveling devices, and chisels were essential to the successful completion of the tasks. A metal workshop with a forge for sharpening tools that quickly became dull would have been crucial, as would other workshops for repairing the wheeled vehicles. Before Anaximander's eyes, then, was a veritable experimental laboratory, where techniques and technologies were invented, developed, refined, and re-tested again and again. As these techniques and technologies were developed and implemented, an understanding of natural laws and of nature's hidden order were revealed and deepened. Ironically, the discovery of these natural laws and principles emerged from a project that was celebrating supernatural powers whose ways were supposed inscrutable or at least not inferable by any human, *rational* process. Religion and science, mythology and rational experimentation and discovery, seem indeed to meet at the Ionian temple building sites. The product of this laboratory was the finished temple, the house of the cosmic power, and the decades-long experiments provided a text book of natural laws for the reflective mind. It is within and against this background of historical and cultural contexts that we must connect the doxographical reports on Anaximander's thought. This task is the purpose of Part I of this book.

B

A SURVEY OF THE KEY TECHNIQUES THAT ANAXIMANDER OBSERVED AT THE ARCHITECT'S BUILDING SITES

In the following chapters of Part I, several problems raised by the doxographical evidence, and debated by classical scholars, will be explored. While many techniques and technologies will be illustrated in each chapter, there is usually one that is pivotal to the clarification or resolution of Anaximander's cosmic thoughts. Let us then, in an anticipatory summary, place those illustrations in front of the reader.

In chapter 2, we will explore Anaximander's identification of the shape and size of the earth with a column drum. Did Anaximander imagine a particular drum, that is, did he imagine the placement of the drum specifically? Keeping in mind that Anaximander held that the earth remains aloft in the center of the cosmos, held up by nothing, did he imagine the drum placed somewhere midway in the column? Did he imagine it as one of the lowest drums in the column, since "lower column diameter" is usually believed to represent the architect's module of the building project? Or did he imagine it as the column base whose measurements are uniform, unlike the drums throughout the column, and for which we have evidence that there were in fact 3 × 1 column bases?

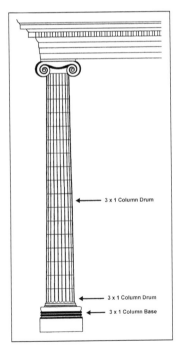

FIGURE 1.1. Where on the column did Anaximander imagine the column-drum earth? (For more detail, see fig. 2.9.)

In chapter 3 we shall explore two problems that seem best to be handled together: (i) What are the cosmic numbers that Anaximander selected to identify the distances to the stars, moon, and sun? (ii) How can the earth be in equilibrium between the cosmic extremes, as Aristotle reports on Anaximander, if the earth is a flat cylinder and not a sphere? The key to these issues is to try to determine *how* Anaximander imagined cosmic structure, and that invites us to explore a question that has never been explicitly raised in the secondary literature: *What point or points of view would have been meaningful to a Greek of the archaic period?* The answer proposed—that the architects routinely distinguished between *plan* and *elevation* views, and so did Anaximander—offers to resolve both problems simultaneously. Anaximander had imagined the cosmos from *both* points of view but the appeal to the *plan* view illustrates the architect's modular approach that shows why the cosmic numbers he selected are 9/10 earth-diameters to the stars, 18/19 to the moon, and 27/28 to the sun, each heavenly wheel being one earth-diameter (= module) in width. By appeal to this architectural technique, we can make better sense about how Anaximander's earth was in equilibrium from the cosmic extremes.

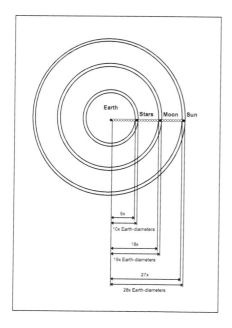

FIGURE 1.2. Anaximander's cosmos in *plan*. (See also fig. 3.5.)

Chapter 4 explores the doxographical reports that Anaximander explained how the fire emanates from the sun, for example, as from the nozzle of a bellows. The light that radiates from the heavenly wheels emerges from an opening, likened to a blow-hole or breathing-hole in the wheel. The metal workshop at the temple building site contained a forge, and the operation of a bellows would be a familiar part of the technology employed there that Anaximander and his Ionian compatriots would have observed routinely. What did an ancient bellows look like? The representations of Hephaistos at the forge show that the bellows was an animal skin, or pair of animal skins, attached to a nozzle which, by pulling up on one skin bag and filling it with air and then pressing it down sharply to expel that air, mimics a breathing animal. After reviewing the archaeological evidence, we will see Anaximander's startling imagination; the hylozoistic cosmos is alive as a fire-breathing living thing. The fire emanates from the heavenly wheels as cosmic exhalation.

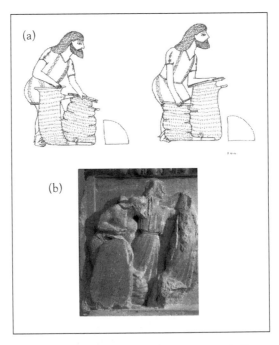

FIGURE 1.3. Hephaistos and the double bellows: Anaximander imagined fiery radiation and cosmic respiration. (a) Reconstruction drawing of Hephaistos working the double bellows. (b) Gigantomachy from the Siphman treasury in Delphi (c. 525 BCE) depicting Hephaistos working the double bellows. (See additional detail in fig. 4.3.)

Chapter five explores the doxographical report that Anaximander imagined the stars, moon, and sun as heavenly *wheels*, each with its rim hollow, and full of fire. Do we have evidence of any hollow-rimmed wheels that might have been the object of Anaximander's projective imagination? After reviewing the evidence of ancient wheel making and a variety of vehicles of transport, a remarkable example presents itself, highlighted by Vitruvius in describing the innovative inventions of the architect Metagenes, builder of the archaic Artemision in Ephesos (fig. 1.4). When this image is entertained, it may be seen as standing in some analogy to an *elevation* view representation of Anaximander's cosmos. Moreover, when we review wheel making and the axle structures with which it is connected, it becomes plausible to consider that Anaximander imagined the cosmos as having an *axis mundi*, and that this axis suggested a cosmos as a great tree, a theme adumbrated by Pherecydes in his earlier prose book of the seventh century BCE.

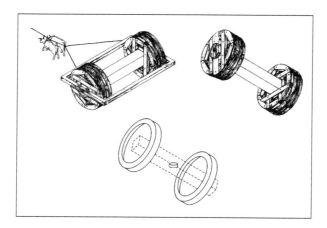

FIGURE 1.4. The architect Metagenes invented a wheeled vehicle with a hollow rim to transport monolithic architraves; Anaximander imagined the sun as a cosmic wheel (summer and winter solstice) connected by an *axis mundi*. (See also fig. 5.15.)

Chapter six explores the *seasonal sundial* with whose invention Anaximander is credited. The sundial provides yet another way to grasp Anaximander's cosmic imagination because it allows us to gain insight into his own solar reflections, the most distant cosmic wheel. So far, the sundial has not been discovered, but we will conjecture how it may have looked, and what the archaeologist might yet find (or has already found and failed to notice for what it was?). The earliest surviving Greek sundials date to Hellenistic times; they are hemispherical or conical. These shapes make sense for those who hold that the earth itself is spherical, but make no sense for Anaximander who believed the earth was flat and cylindrical. Moreover, in the doxographical testimonies about the sundial, there are also reports that Anaximander invented or introduced the gnomon, made a map of the earth, and created a cosmic model. We will start by setting a gnomon in the center of a column drum and explore how the seasonal markings of the sun's shadow might well produce a series of lines that remarkably resemble the architect's column drum, exhibiting *anathyrôsis* and *empolion*, installation techniques that secured lateral and vertical fixity. Moreover, we shall conjecture about how it might have been that Anaximander set up on a column drum in Sparta the *same* gnomon he used in Miletos and, by virtue of the shadows cast, made his earthly map and model using just this drum. The conjecture is that they all might well have been one and the same thing—a seasonal sundial, map, and model (fig. 1.5).

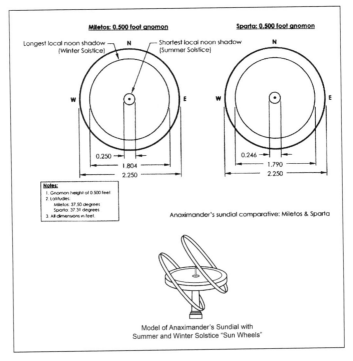

FIGURE 1.5. A possible reconstruction of Anaximander's *seasonal* sundial, set up as a model in Sparta. (See also fig. 6.11.)

C

AN IMAGINATIVE VISIT TO AN
ANCIENT GREEK BUILDING SITE

To reflect more deeply on these matters we shall try to re-create a visit to the ancient temple building site. We shall follow the outlines relevant to our case thanks to the genius of archaeologist and architect Manolis Korres in his *From Pentelikon to the Parthenon.* While it is true that Korres' splendid study is on a building program in Athens and not Ionia, and that it traces the fate of a column capital from the quarry to the building site in the first part of the fifth century BCE, and not more than a half-century earlier in Ionia as is our focus, it is not implausible that key technological ingredients were largely the same. By imaginatively thinking through a range of techniques that Anaximander would have witnessed at the temple building sites and the successful results achieved after a long process of trial-and-error experimentation, we begin to establish the intellectual ground plan that will enable us to reenact Anaximander's past thoughts more clearly still. The facts of this matter do not exist independently of us; we make the facts by thinking through the historical and cultural context of temple building and connecting them to the surviving doxographical fragments.

Before the temple took shape, the scaffolding was already set up around the proposed building.[17] The scaffolding provided a *frame* for the project; it was indispensable to the successful construction itself, of course, but it is plausible that it created an image with long-lasting impact for Anaximander and his community. Anaximander was credited with making the first Greek map of the inhabited earth. Unlike the earlier senses of mapping that we find, for instance, in Homer, when Odysseus is instructed to keep Chios on his "right hand" as he passes, the successful making of a map needs a frame, as Heidel pointed out almost a century ago, to contain the whole region in question.[18] The temple was the house of the god; it was the god's cosmos writ small. To create that cosmos, the architects began with a frame. The scaffolding provided a frame for the cosmos to come (fig. 1.6).

There were a host of problems at the quarry. The quarry foreman, and the architect, must have been acutely aware of issues dealing with the stone lying in the quarry. Time after time, work that began on a piece of stone had to cease because imperfections—natural fissures and cracks—were discovered lying below the surface. Therefore, in order to avoid ruinous delays and uncontainable costs, the architect had to know intimately both the nature of the stonework and this quarry in particular. The work of removing the selected stone from the parent rock was carried out by time-tested techniques practiced by the Egyptians, and elsewhere, using sockets, wedges, and levers (fig. 1.7). In the case of this mass of stone intended to become a column cap-

FIGURE 1.6. Scaffolding around the Athenian Parthenon.

FIGURE 1.7. Working in the quarry.

ital in the pre-Persian Parthenon, one estimate placed the job as requiring five days to carve deep grooves around the stone, with twenty-one sockets for wedges.[19] The iron wedges are fitted into the sockets, sandwiched between iron splints. They are pounded into place by iron hammers, each weighing perhaps a talent (a talent being equivalent to roughly twenty-six kilograms).[20] As the pounding of the hammers continued, the block of stone finally began to give way, and the clue to this moment was announced by a change in the sound of the creaking, a sound that the expert quarryman and architect knew very well after much experience. Since the metal tools quickly became dull from hard use, a nearby metal workshop with a forge would have been visited frequently to repair the worn tools.

Then the quarry foreman had the problem of removing the stone from the quarry, often requiring that the stone be moved up a steep incline. Next, the stone would have to be removed from its original location to a platform from which it could be pulled out of the quarry. Such a task required rope, pulley, winches, levers, wooden beams, and rollers (fig. 1.8). This particular piece of stone, it has been estimated, would likely have weighed some 1,600 talents, or

FIGURE 1.8. Removing blocks from the quarry.

forty tons, before the excess material had been cut away.[21] The column capital itself weighed some five hundred talents, and that means that the excess material weighed on the order of twenty-two tons. Naturally, the quarry foreman and architect had to plan with exceeding care to make sure as much useable material was protected, and this added yet another difficulty to the task. The time it took to quarry the stone and half-finish the capital has been estimated to be on the order of two months,[22] and when we keep in mind the dozens of capitals needed for the external colonnade, and that these capitals were only part of the enormous work that was required for building the temple, some sense of this extraordinary enterprise becomes clearer.

From the quarry, the architect had to provide the means for transporting the blocks to the building site (fig. 1.9). This task required special vehicles of

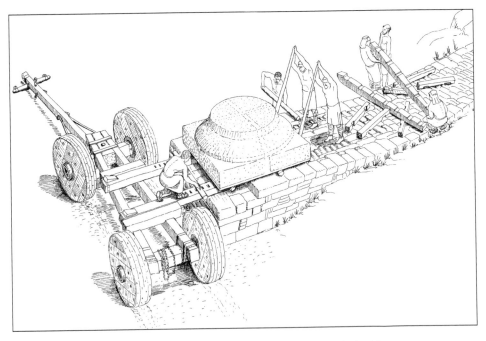

FIGURE 1.9. Transporting blocks from the quarry to the building site.

transport because the load was so much greater than that of common tasks. We can begin to acknowledge the Ionians' appreciation of truly circular-shaped wheels, and the role of the axle in bearing enormous loads.

A variety of techniques, in addition to the use of winches, were called for to achieve the task of moving heavy stones up an incline. Sometimes ingenious solutions employing the simplest technologies sufficed; in this case, blocks could be delivered up a steep incline when a team of oxen pulled a rope-drawn cart, circling a heavy post (fig. 1.10).

A visitor to the building site could also witness other kinds of lifting devices that depend on the original scaffolding for support, and we see the techniques for column-drum installation with the bosses still intact. Ropes would have been placed around the four bosses, and the winch–pulley devices would hoist the drums into place (fig. 1.11). The bosses would have been removed in the final stages of installation.

The proof that there were comparable lifting devices in use during the earlier archaic constructions is provided by the example in figure 1.12 of an unfinished column drum exhibiting a boss, from Artemision C in Ephesos, dating to the mid-sixth century BCE.

FIGURE 1.10. Transporting stones up an incline by means of a team of oxen that pulled a rope-drawn cart, circling a heavy post.

FIGURE 1.11. Lifting devices for installing column drums.

FIGURE 1.12. An unfinished column drum with boss, from the archaic temple to Artemis in Ephesos, dating to the mid-sixth century BCE.

Other lifting devices have also been proposed, including truly enormous machines (fig. 1.13), the kind alluded to by Aristophanes in the *Clouds*, where Socrates was hoisted above the "Thought School" in a crane.

Let us reflect on one more general problem of construction, the fashioning of each architectural element—here the preparation of the column capital. In this case, square blocks from the quarry would be geometrically measured, and then, with the use of caliper (i.e., compass) and straight edge, the circular form would be carved (fig. 1.14).

The central point to emerge from our brief and selective visit to an ancient temple building project, and the presentation of some techniques found there, is to remind the reader of vast enterprises in applied geometry that were required for the construction of monumental stone temples. The

FIGURE 1.13. Reconstruction of the great crane used to complete the installation of the upper orders of the temple.

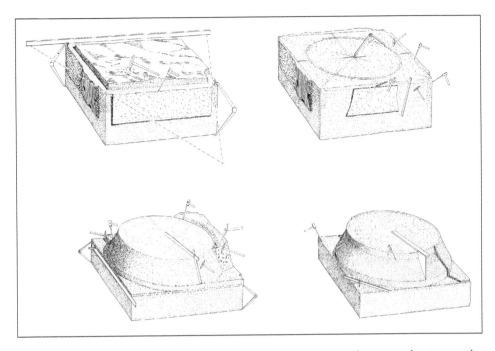

FIGURE 1.14. The stages of geometrical imagination in preparing a column capital, using straight edge and caliper.

technologies and array of techniques provided a veritable experimental laboratory for reflection by a receptive mind. Anaximander imagined the cosmos—the house in which we all live—and its construction, as analogous to the architecture of the house of the cosmic powers. He had before him these many activities to watch carefully, and many more techniques not pictured here, that informed his cosmic reflections. It is archaeological resources and artifacts that supply the details of this world of activities in which his thought was nurtured.

D

ARCHITECTURAL PLANNING

There is one more theme that must be mentioned in any discussion of a visit to the building site, and this concerns techniques for architectural planning. In my earlier books I have discussed those techniques at great length and provided the evidence for them; the reader may be directed to them for much fuller details.[23] Here I want to emphasize that the architect had to imagine these enormous buildings from more than one point of view. When we assess

29

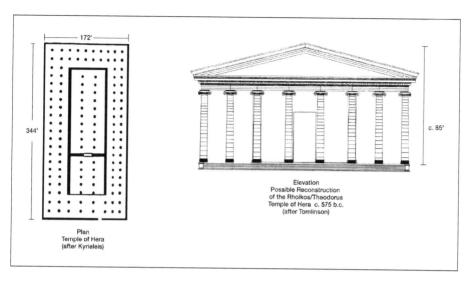

FIGURE 1.15. The archaic Temple of Hera in Samos (Dipteros I), c. 575 BCE, in Plan (l) and in Elevation (r).

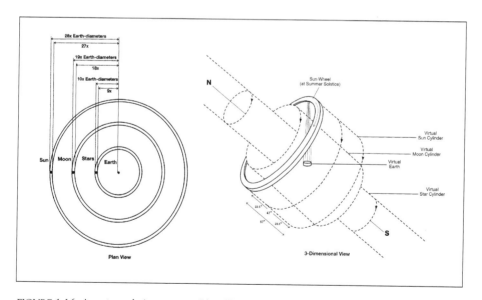

FIGURE 1.16. Anaximander's cosmos in Plan (l) and Elevation (r).

the intricacies of the measurements that the buildings exhibit, it is clear that the architects worked from a written plan. It has been doubted that the Greek architects had scale models, but clearly there must have been some sort of informal sketches with numbers and measures identified (I go through some of these considerations in chapters 2 and 3).

In order to control the gigantic building as it was built, the architect had to imagine each course of stones, each level, one upon the other. This requirement forced the architect to imagine the building in *plan*, that is, literally from a vantage point 90 degrees above. Of course in the ancient world, unlike our hovering helicopter, there is no way for someone to actually have that experience. Thus, the architect had to engage in an imaginative projection. The imagining or viewing the building from 90 degrees above is a *plan* view. Among the other orthographic projections other than plan that the architect had to command we might generally refer to as *elevation* views. And in order to appreciate how very different an object appears from these two kinds of view, in figure 1.15 I present a plan view of the archaic temple of Hera in Samos, and next to it a reconstruction of the same temple in elevation. And in figure 1.16 I present an illustration of Anaximander's cosmos in plan and in elevation. It is from the architects' projects that he came to imagine the cosmos from two very different points of view.

It is clear that Anaximander did not simply imitate the architects but rather adopted and adapted their technique because he came to imagine the cosmos as architecture built in stages.

TWO

Anaximander's Cosmic Picture

The Shape and Size of the Earth

A

THE DOXOGRAPHICAL REPORTS

Τὸ δὲ σχῆμα αὐτῆς (sc. τῆς γῆς) γυρόν, στρογγύλον, κίονος λίθῳ παραπλήσιον.

—Hippolytus, *Ref.* I.6,3 [D-K 12A11.3]

The shape of the earth is concave, rounded, similar to the drum of a column.

ὑπάρχειν δέ φησι τῷ μὲν σχήματι τὴν γῆν κυλινδροειδῆ, ἔχειν δὲ τοσοῦτον βάθος ὅσον ἂν εἴη τρίτον πρὸς τὸ πλάτος.

—Pseudo-Plutarch, *Strom* 2 [D-K 12A10]

He [sc. Anaximander] says that the shape of the earth is cylindrical, and that it has a depth that is one-third of its width.

Ἀ. λίθῳ κίονι τὴν γῆν προσφερῆ.

—Aetius III, 10,2 [D-K 12B5]

The earth resembles a stone column.

B

THE SCHOLARLY DEBATES OVER
THE TEXT AND ITS INTERPRETATION

AFTER SURVEYING the scholarly debates concerning the size and shape of Anaximander's earth,[1] it becomes clear that there has been no doubt that

Anaximander identified the shape with a cylindrical form, likening it to a column drum. Moreover, the general consensus has been that this flat and cylindrical earth was wider than it was deep. But does the debate take us to the temple building site? Or was the allusion to a column drum just a throwaway, an illustrative comment made only to clarify the hypothesis that the earth was a flat disk? The commentators allude, at times, to some facts about column drums as they try to clarify key terms in Hippolytus' doxographical report, namely, *guron*[2] and *strongulon*.[3] Guthrie, Kirk-Raven, Robinson, and many others indicate an awareness of this architectural element, but explore no further.

The ongoing debates try to clarify the shape and size of the earth. The discussions focus on the manuscript reading of Hippolytus; the text reads: "the shape of the earth is watery, rounded, like a snowy stone."[4] As Kirk-Raven observed the situation, Roeper had proposed the emendation of "curved" (*guron*)[5] for the impossible "watery" (*hugron*).[6] Kirk-Raven adopted Roeper's suggestion and consequently took "rounded" (*strongulon*)[7] as a gloss, since *guron* (γυρόν) is construed to mean "curved" and *strongulon* (στρογγύλον) is taken to mean "round." The manuscript had another problem, namely, that "snowy" (*chioni*),[8] not "column" (*kioni*),[9] appeared. Kirk-Raven emended *chioni* to *kionos exempli gratia* since not only Hippolytus but also Aetius supplies "stone column."[10] Thus, the shape of the earth is not "watery" and it is not somehow analogous to a "snowy" stone. Instead, the shape of Anaximander's earth is curved, rounded, and likened to a stone column.

There has been debate over the construe of *strongulê*[11] in a variety of conflicting doxographical reports describing the *spherical* shape of the earth for Democritus and Parmenides, and elaborated upon in Plato's *Phaedo*.[12] Heidel pointed out, however, that *strongulê* was not used even generally for a "sphere" in the fifth century,[13] but rather to describe a cylindrical shape. Theophrastus in his *History of Plants* describes unsquared logs as *xula strongula*,[14] and in another place he refers to a plant having a "round stalk" as *strongulokaulon*.[15] Burnet pointed out that the same term is used to describe the earth in Diogenes of Apollonia, whose conception was almost certainly disk-shaped.[16] Thus, in several cases that we do have, including those of Parmenides and Democritus, the term *strongulê*[17] is used to refer to something cylindrical, and thus describes the circumference shape. Robinson had astutely noted that there was perhaps a linkage to column drums in a Greek column where the drums had to be made concave and convex respectively in order to provide proper seating when the drums were placed on top of one another.[18] While he does not say so explicitly, he makes it clear that the meaning of *guron* is taken to mean "convex/concave" in contradistinction to *strongulon*. Had Robinson followed the lead to the architectural technologies of the temple builders, the appeal to both terms might have suggested that Anaximander had firsthand knowledge of them, and implied a deeper interconnection to the community of architects and their projects. Robinson followed this analogy no further.

According to scholarly consensus, the testimonies that survive show that Anaximander believed that the shape of the earth was flat and cylindrical, somehow analogous to a stone column. Diels had taken the expression "the earth resembles a stone column"[19] to be genuinely Anaximander's, and this technical analogy between the shape of the earth and an architectural element has never been the subject of serious debate. Of course, the expression "stone column" conjures a range of possibilities, including a Doric or Ionic column, but this identification is immediately ruled out by appeal to the testimony of Pseudo-Plutarch, who in the *Stromateis*, while identifying the shape of Anaximander's earth with a cylindrical form, provides the proportions of 3 × 1 to describe its size.[20] The expression he offers is *exein de tosouton bathos hoson an eiê triton pros to platos*.[21] The natural reading of this report should be "to have as much depth as would be a third part (*meros*) in relation to the width," that is, to have a width of 3 and depth of 1, though Martin interpreted the expression to mean to have a height that is three times the diameter.[22] If the argument is that Anaximander's earth is held in the center *dia tên homoian pantôn apostasin*,[23] then the proportions of 1:3 as opposed 3:1 should not count as an objection. Martin's rendering is not impossible, but the vast majority of scholars regard his construal as unlikely: Anaximander's earth is a flat disk, 3:1.[24] The strength of the case for a 3:1 earth has also rested on the appeal to the tradition of Homer and Hesiod, who conceived the world as a flat disk. And nothing we know about Thales or Anaximander suggests that he held a different view from these predecessors.

Can archaeology substantiate or undermine these proposed emendations? Can archaeology shed deeper illumination on Anaximander's belief system as we wrestle with the surviving doxographies? We shall begin with the architectural element and shall venture farther along this line that takes us, ultimately, to the temple building sites where we find Anaximander and his audience. There we are left with a new architectural question for which only archaeological clues can help us explore possible answers: Was the exemplar for Anaximander's cosmic, analogical, and imaginative reflections a drum situated somewhere in the column, or was it a column base? Can a plausible case be made that Anaximander imagined the earth's shape in terms of a drum situated in a particular place in the column construction?

C

THE ARCHAEOLOGICAL EVIDENCE

For Anaximander, the shape of the earth is likened (a) to a stone column, (b) its proportions are 3 × 1, and (c) its shape is further explained as "curved and round." What is clear at any rate is that a *piece* of the column is the focus of the analogy. Can the archaeologist shed light on the *piece* of the stone column that Anaximander presumably had in mind? What more can the archaeologist tell us about column construction that would clarify the doxographical report?

To explore this matter, we turn first to consider the art historical evidence. When we do this, we shall understand better the evidence for the symbolic meaning of the column and its appearance as a drum or base. Afterwards, we shall return to the evidence from the temple architecture to focus in on the architectural elements of column drum and column base.

C.1

THE COLUMN AND ITS SYMBOLIC
FUNCTION IN ARCHAIC GREECE

Undoubtedly, the column and its parts had symbolic meanings for the ancient Greeks. The problem is to determine more precisely what those meanings were. Because vase painting is a highly conventionalized medium, art historians argue, curiously enough, that a "column" is not always a "column." In the relatively small space suitable for a painted image, it is not possible to paint in all the details of a scene or story, and so individual iconographic elements must be seen as shorthand for larger narratives. Kathleen Lynch has argued recently that the appearance of a column or column drum or column capital serves as such shorthand to indicate either the presence of the built environment or to mark a transition from one scene to another in a vase painting.[25] These background considerations need to be kept in mind as we examine the specific themes in which columns appear on archaic vases. For, of course, sometimes a column *is* just a column.[26]

The Lakonian Hunt Painter cup (ca. 565–550 BCE), also referenced as the "Samos cup," is commonly believed to depict Sisyphus in the underworld. Consider the fragment from the cup shown in figure 2.1. At the center of the debate is a single Doric column that seems to lie underneath a hill. There is no Doric entablature and the roof appears to rise straight up from the ground level. The proportions for the column shown here are normal for a full Doric column. It has been suggested by Schaus that, following Anaximander's theory, the introduction of a column identifies the location of the scene as the underworld.[27] And the specific shape of the stone that Sisyphus rolls up the hill is strikingly suggestive of a prepared column drum.

A similar vase, one by the Kleophrades Painter (ca. 500 BCE), now in St. Louis, also displays a column in its pictoral telling of the myth of Sisyphus (fig. 2.2). In this case, Sisyphus is depicted with Hermes and Cerberus. Just behind and above the stone he bears is the upper part of a Doric column with no connecting architectural features. Scholars who suggest some connection

FIGURE 2.1. Lakonian Hunt Painter cup (c. 565–550 BCE), Sisyphus in the under-world, with a column. The "Samos Cup."

of Anaximander's theorizing with the vases tend to rely on the fact that a piece of the column is presented without any other architectural structures. In this instance, the column is short, and Schaus also suggested that it is con-sistent with Pseudo-Plutarch's testimony that the earth's cylindrical form has a depth a third of its width.[28]

Much of the argument for the representation of Sisyphus on the Samos cup relies on the interpretation of the 'hill' and the column beneath it. The column is important in the scene, but serves no architectural function. It is a single Doric column with no Doric entablature and it stands alone. The appearance of the column on the St. Louis vase, which clearly depicts the underworld, forms the main argument. As Schaus puts it,

> it seems likely that the columns seen in both the Samos and St. Louis vases are meant to distinguish the scene as the punishment of Sisyphus, perhaps simply by localizing it in the Underworld. But what exactly may the column

FIGURE 2.2. Kleophrades Painter (c. 500 BCE), Sisyphus with Hermes and Cerberus, with a column. [St. Louis]

stand for here in order for it to carry out this supposed function? The intriguing possibility exists that the vase painter had in mind the theories of Anaximander who taught that the earth was actually shaped like a column. In this interpretation the hill of Sisyphus, standing directly above (as on the Samos cup) or directly below the column (as on the St. Louis amphora), is understood to be in the Underworld, situated at the end of the column-shaped Earth opposite to that on which mortal men dwell.[29]

The inclusion of a column without any connecting architectural features in mid-sixth-century BCE. Lakonian pottery has been seen as evidence in support of the tradition that Anaximander spent some time visiting Sparta, in addition to the two doxographical reports that indicate that Anaximander set up a seasonal sundial in Sparta[30] and advised the Spartans of precautions in case of earthquake.[31] Further evidence for Anaximander's influence in Sparta has been argued by Yalouris,[32] referring to a work of the Lakonian Arkesilas Painter (ca. 550) now found in the Vatican Museum (fig. 2.3). Yalouris references the Arkesilas Painter's vase, which has a representation of Atlas and Prometheus where the entire scene is seen to be supported by the upper part of a Doric column placed in the exergue.[33] The column is connected to *no* architectural features; rather, it connects to the exergue as if supporting the ground-line that both Prometheus and Atlas stand on. Further, he calls to attention the fact that details in the scene soundly conform with the philosophical concepts of the cosmos at this time. The two figures represent the two boundaries of the world; in the East at Caucasos (right) Prometheus is bound to a column, and in the West (left) Atlas supports the heavens. It is possible, even, that the column to which Prometheus is bound is an 'echo of the column-shaped earth.' Schaus argues for Anaximandrean influence; the painter used the Milesian concept of a starry heaven as a sphere surrounding the earth in order to represent the sky.[34] But, as I have argued elsewhere, the idea of a spherical heaven is not Anaximander's. Anaximander's cosmos is cylindrical, modeled on the imagery of a great cosmic tree; the original fire surrounded all like bark (*phloios*)[35] around a tree (*dendron*).[36] In any case, with this vase, Hans Jucker was the first to identify the cosmological ideas of Anaximander in Lakonian pottery.[37] These points were again argued by Gelzer[38] and Yalouris, and both added to the discussion a reminder that Anaximander believed the earth was column-shaped with the opposite ends being *inhabited surfaces*: "on one of its flat surfaces we walk, and on its opposite side there is the other." (*tôn de epipedôn hô men epibebêkamen, ho de antitheton hyparchei*).[39] Based on this art historical exegesis of the archaeological evidence, it is plausible to suggest that these column representations depict the two sides of the earth. The column in the underworld shows life, both mortal and divine, on either end. The art historical and archaeological evidence, then, might be seen to lend

FIGURE 2.3. Lakonian Arkesilas Painter (c. 550 BCE), Atlas and Prometheus supported by the upper part of a column.

support to the view that Anaximander might well have maintained the existence of the antipodes, walking on the other side of the column-drum earth. And this would suggest also that Anaximander embraced a conception of space where long-believed oppositions such as "up" and "down," "left" and "right" were no longer to be accepted.[40] Anaximander's cosmological conceptions introduce new ways of thinking about space; there is depth in space where the sun lies behind the moon, and the moon behind the stars, which themselves are already very far from us. No more is there a crystalline sphere in which the heavenly lights are fixed, or nailed into, an older conception to which his younger contemporary Anaximenes would soon return. The archaic artist delivers an idea in pictorial form, replete with its shorthand technique of narrative; for the artist, the column representation can separate the mortal from the heavenly or the mortal from the chthonic. We can be sure that the column had a variety of symbolic meanings for the Greeks of archaic times.

Differences in both the time and place of construction might account for why the Samos cup shows the column beneath the hill and the St. Louis amphora depicts it above. Or it might also be that either the Kleophrades Painter was less knowledgeable of the reports of Anaximander's opinion, or the conventions to represent that opinion were not established. After all, if one considers for the moment the difficulties of imagining the cosmos itself, we now have the further problem, in artistic shorthand, of representing it in a simple and recognizable way. And from which point or points of view would such representations be meaningful to a Greek in the archaic period? The Samos cup was made circa. 565–550 in Sparta while the St. Louis amphora was made circa. 500 in Athens; do these representations stand in continuity or discontinuity? Another idea, as espoused by Schaus, is that "the Kleophrades Painter (St. Louis amphora) may also have had difficulty in grasping Anaximander's idea that there is no up or down when the Earth is considered the center of the universe."[41] Another reason the Samos cup may have the column placed under the hill is to emphasize that Hades is on the opposite side of the earth from the surface where living men are. For Schaus, the Samos cup is more true to Anaximander.[42]

What can we conclude from this discussion? Two lines of interpretation can be identified: The first is about archaeological and art historical infer-ences;[43] the second concerns the cultural context in which Anaximander's thought unfolded.

A reflection on both Schaus and Yalouris shows how art historical *infer-ences* have been made, given archaeological artifacts. Both Schaus and Yalouris were trying to illuminate the materials on vase paintings. They observed the appearances of columns and parts of columns and then reflected on what we already know, thanks to classical scholarship on Anax-imander. The results are a series of plausible inferences suggesting one of three possibilities: (1) Anaximander's ideas had sufficient outreach within the archaic community to inspire the vase painters, or (2) current discus-sions within the archaic community inspired the painters who in turn inspired Anaximander's expressions, or (3) at all events, without specifying the causal direction of inspiration, the column and its parts already had established cosmic significance; within this cultural *Weltanschauung*, Anax-imander's ideas flourished. This study, however, while beginning with these resources, proposes a different line of approach; for that reason I have sug-gested we might better term the method a kind of *inverse archaeology*. Given the philological evidence collected from the tertiary reports, we must ask the archaeologist to provide evidence that corresponds broadly. We must ask fur-ther: What is the evidence for archaic columns and drums, and the tech-nologies by which they were produced? Neither Schaus nor Yalouris made the exegesis of Anaximander's belief system their central objective; this, however, is central in our study.

The second line of interpretations—concerning the cultural context—follows from these considerations. Column drum construction was new to the archaic communities in the early sixth century BCE. The community marveled at the *thaumata* created by means of this new technology.[44] It is not implausible that in the mid-sixth century cosmic speculations about the Earth and the column drum were sufficiently well-known to find their pictorial expressions on vases; they were original enough to capture the imagination of the artist whether through Anaximander's influence directly or from the temple architects themselves. Thus, the thesis of a column-drum–shaped Earth was received by a community for whom a range of symbolic meanings of the column were already current.

C.2

ARCHAEOLOGICAL EVIDENCE FOR ARCHAIC COLUMN CONSTRUCTION

Archaeologists have helped us to understand that a major change in temple building was inaugurated in eastern Greece in the archaic period, and these projects offer us valuable insights about the cultural context and its technological innovations.[45] Could Anaximander's description of the earth have referred more specifically to something that he, and those in the community who might likely have been his audience, saw at the building sites? The provisional answer is most certainly "yes," but there is more than one candidate to fit the proportional description. Was the architectural element—the part of the column—that Anaximander had in mind a "column drum" or "column base"? To grasp the difference is in part to grasp the particular symbolic meaning that Anaximander selected to convey.

In eastern Greece, the transition from wood to stone architecture became pronounced in the first half of the sixth century BCE, although preliminary stages can be detected already in the seventh century.[46] Sometime before 575 BCE a decision was made in Samos to erect an enormous temple to Hera, with columns made completely of stone. This was the so-called Theodoros temple, usually referred to as Samian Dipteros I. Not much later, around 560, the construction process began to create enormous stone temples in Ephesos to Artemis and in Didyma to Apollo, the neighboring rivals of Samos. Both the Samian Dipteros I, dedicated to Hera, and the archaic Didymaion, dedicated to Apollo, were made predominantly of limestone. In Ephesos, the archaic Artemision, dedicated to Artemis, was constructed of marble, and became known as one of the seven wonders of the ancient world. In all three cases, the transitions from wood to stone architecture brought with them a host of technical problems, though the work in Samos seems to have led the way.

In earlier temple building in Ionia, available timber, sometimes imported from Lebanon or elsewhere in Anatolia, was used as columns to hold up the roofs of the temples. Through the course of the seventh century, the Samian temple to Hera was enlarged more than once, but this was achieved by lengthening the temple, not widening it; thus, the problems of carrying the load of the roof were still manageable by the available timbers. However, when a decision was made sometime in the sixth century BCE to achieve greater monumentality, the new plan required that the buildings would have to be widened. Unfortunately, the available timber was no longer sufficient to support the weight of the roof. It was at this time that the roof, commonly made of thatch, was replaced by more permanent material, namely, terracotta tiles, sometimes held in place on beds of clay. But, in any case, the larger, wider roof design flattened out to prevent the tiles from sliding, and this too contributed significantly to load-bearing problems for the roof.[47] Stone columns, then, were a solution to the problem of carrying the load of the heavier roofs. To achieve the monumentality to which the archaic architects aspired, the columns were also made much higher than before, stretching to fifty or sixty feet. While the available timbers could not carry the load, the delivery of monolithic columns from the quarries to the building sites, sometimes many kilometers away, created new challenges for the architects. Since the dipteral temple designs, as the archaeologists' reconstructions have shown, required more than a hundred columns to fill the colonnades, the architects were confronted with a daunting task of quarrying and safely transporting an exorbitant number of monolithic columns. At full height, each column would have weighed tens of thousands of pounds, and the difficulties of this production and delivery of monolithic blocks were insurmountable. Given this challenge, the architects initiated an invention new to Ionia, the production of column drums.

Archaeologists have been able to help us discover the range of proportions that were suitable for column drums and bases in these monumental temples. The evidence ranges broadly from drums and bases 2×1 to 5×1, and drums and bases exhibiting the proportions of 3×1 were plentiful. Thus, Anaximander sought to explain to his audience the size and shape of the earth by analogy with an ingenious innovation in temple architecture; he would have found a community intimately familiar with column drums and bases, and the proportions 3×1 would have been immediately familiar and appropriate.

The column became the defining feature of the Greek temple, and we have already shown that there is archaeological and art historical evidence to support the claim that the column had symbolic significance. Now, we turn to the technical problems of column construction that captured the attention of the architects and the communities of eastern Greece, to consider whether the "column drum" or "column base" proves to be the more likely exemplar for Anaximander's model.

The solution to the problem of quarrying, transporting, and installing columns was the innovation of column-drum construction. More than a thousand drums would have had to be produced to complete the temple colonnades, and of course it was much easier to deliver to the building sites the smaller drum blocks. But, while the delivery of column drums was greatly facilitated by this innovation, a new and extremely difficult problem was created at the same time—that of installing the drums. It is fair to say that this problem was more difficult than quarrying and transporting the drums. In order to get each drum, finished at the site prior to installation, to sit securely on the preceding drum or base, two time-consuming techniques had to be performed to achieve a successful result. The first part we refer to by the modern term *anathyrôsis*.[48] Instead of requiring the entire surface of the drum to sit on the previously installed drum, as a labor-saving device the load-bearing surface was restricted to a band running around the circumference of the drum face (like the band around a door, a *thyra*). That circumference band had to be carved with great care to be sure that it was both uniform in distance from the center and had attained exactly the same leveling over its course; this was no small challenge since the drum faces were often as large as two meters in diameter. Otherwise, the architect could not control that the drum would sit perfectly upon the lower drum and still have the stability to carry the weight of the upper orders of the building. Thus, the preparation of the drum face would have had to be *both strongulon*, round at the circumference, and *guron*, concave, between the circumference band and the drum's center, so that none of the inner part of the drum or base protruded, thereby challenging the stability of the seating of one stone on top of the other. The archaeologist's display of the ancient architect's technique of *anathyrôsis*, then, offers to resolve the textual problem about the correction to the manuscript reading in Hippolytus and, moreover, shows that *strongulon* was *not* a gloss of the corrected *guron* but instead referred to a different aspect of the earthly shape.[49] In this way, archaeology shows its usefulness to our understanding of some aspects of Anaximander's thought and helps us understand *how* he thought.

Consider the evidence for *anathyrôsis* exhibited on an archaic drum fragment, with round *empolion*, from the archaic Didymaion, and next to it a surviving example of an archaic drum from the Samian Dipteros II temple in figure 2.4.

In addition to *anathyrôsis*, the other part of the technology of column-drum construction is called the *empolion*. The *empolion* is a dowel, usually of wood, either rectangular, square, or round in shape. To use it, a comparably shaped hole must be carved into each drum exactly at its center. The dowel is then inserted into the drum so that it may be connected to the hole carved into the center of the other drum to which it is being connected. Then, by means of a lifting device with ropes tied around the four protruding bosses, the drum would be lowered into place. This technique allowed the new drum to be seated precisely and thus avoid chipping the stones as they came into

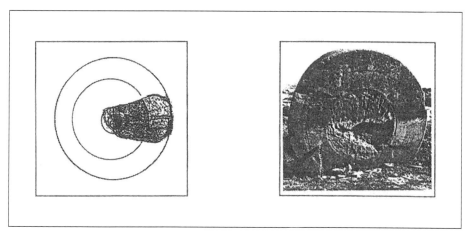

FIGURE 2.4. Drum fragment from the archaic Didymaion exhibiting *anathyrôsis* and a round *empolion* (l), and next to it (r) a drum exhibiting *anathyrôsis* found in Samos from Dipteros II.

contact with one another. Figure 2.5 is an illustration, after Orlandos, that displays how the *empolion* was used.

Now, the last part of this technical discussion is to decide whether the identification of a 3 × 1 column piece should be made with a column drum or a column base. Although I have been arguing that a column drum is more likely the reference for Anaximander's cosmic speculations about the shape and size of the earth,[50] the suggestion has been made recently that a 3 × 1 column *base* might more likely have been the image that transfixed Anaximander's mind.[51] The archaeological evidence shows that, while early column bases were not always 3 × 1, there were several prominent examples that were. Let us explore this argument.

In a recent excavation in Didyma, a column base was discovered; the diameter is 118.5 cm, and the height is 37.5 cm. The proportion is nearly 3:1 (exactly 3:0.95). This votive offering base, or *Weihgeschenkbasis* as the German archaeologists refer to it, was found in the northwest part of Didyma in 1991 during the excavations of the Sacred Road connecting Didyma with its patron city of Miletos. It is a column base of marble lying in situ on the east side of the Sacred Road and belongs to a double agalma of two columns with Ionic capitals, both of which were found in a deposit beside the base. The second base is not preserved. One part of the column (diameter ca. 50.7 cm) was also found. The capitals can be dated to the first half of the sixth century BCE by stylistic evidence.[52]

There is another cylindrical base in Didyma made of limestone and now lying near the round altar in front of the temple to Apollo. The diameter of 106.2 cm and height of 31.6 cm produce a proportion of nearly 3:1 (exact:

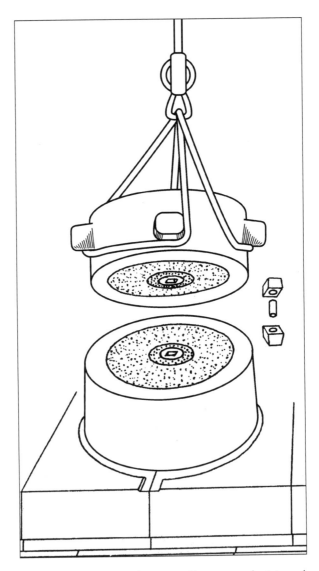

FIGURE 2.5. Column-drum installation—*anathyrôsis* and *empolion* are exhibited as well as drum bosses and use of rope with lifting device.

3:0.89). This column base is not exceptional, but rather quite similar to the other early bases we know of, dating to the seventh and first half of the sixth centuries BCE. Figure 2.6 depicts this cylindrical base.[53]

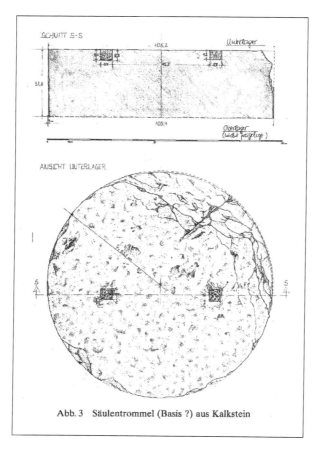

SCHNITT S-S
106,2
31,6
105,4
ANSICHT UNTERLAGER
Abb. 3 Säulentrommel (Basis ?) aus Kalkstein

FIGURE 2.6. Archaic column base, 3 × 1, from the temple of Apollo at Didyma.

The general development of Ionic columns and their bases has already been presented by Gruben.[54] The fact is that, up until this point, the oldest known column bases of the seventh and first half of the sixth century display a purely cylindrical form. These Ionic column bases have been discovered in Delos and in the Oikos of the Naxians (beginning of the sixth century BCE) in the proportions of 3:1. Contemporaneously, in Delphi, we have a column base for the column of the Naxians in the proportion of 2:1 (dating to about

570 BCE), as is the so-called column in Aegina, dating also to the beginning of the sixth century.[55] Column bases dating to the archaic period have also been discovered in Naxos, at the temple of Sangri, and again with 2:1 proportions.[56] An example of an Ionic column with a 3 × 1 column base appears in figure 2.7, after Gruben.

When we consider the archaeologist's evidence, we see that Ionic column bases were of a distinctive and cylindrical form right from their earliest appearance in archaic times. While all the early examples of Ionic bases do not exactly display the defined proportions testified to in the doxographical reports of Anaximander's earth, some of them indeed do. We find this proportion at least a few times among the early examples of cylindrical bases, while others tend to have the proportion 2:1. So, no doubt, the examples of early column bases suggest that it might well have been that image—the column *base*—that Anaximander had in mind when he imagined the shape of the earth in terms of an architectural element.

Those who wish to argue that a column base is more likely the metaphorical image that burgeoned in Anaximander's cosmic speculations might emphasize the drawbacks from identifying the image with a column drum. For it may be doubted whether any single drum as a small and absolutely secondary part of the column shaft could instead have served as Anaximander's "image" for the center of the world. The argument on behalf of identifying an architectural element with Anaximander's imagination—the column *base*—turns out to be an argument against the identification with a column *drum* somewhere in the column. To see what is at stake, let us consider the wider argument.

To argue instead that a column drum was the more likely identification for Anaximander's imagination, we must be clearer yet about the arguments for the column-*base* theory. From the point of view of the architect, the column consisted of three parts: a column base, a shaft, and a crowning capital. Thus, it may be objected that because the drums have a sense only *in pluralis* and their form exists only during the process of making the stone column—because upon installation the drums were to become "invisible," leaving only an apparently monolithic column to appear—the column *drum* is a less likely selection for Anaximander's imagination. On this view, the drums as "drums" no longer exist as independent tectonic elements once they have been incorporated in the column. Moreover, column drums never have metrologically defined proportions. This is because the column construction displays *entasis*, the tapering of the column as it reaches upward to the capital. Thus, the column-drum diameters diminish as each drum is placed atop the previous drum, and there is no uniformity of column diameter throughout the column. Moreover, there is no metrologically defined height for each drum either since the individual drum depends on the available stone material, which was used as economically as possible. Thus, the only requirement for the architect in

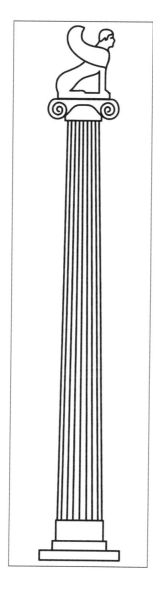

FIGURE 2.7.
Reconstruction
of a standing
Ionic column
with 3 × 1 base,
after Gruben.

planning the height of each drum was to be sure that the composite of drums in each and all the columns, finally, reached exactly the same height for the placement of the uniformly sized capitals. Thus, the argument for identifying the 3 × 1 proportions with a column base rather than a drum for Anaximander's imagination rests on the enormous *tectonic* and also *symbolic* meaning of this early form of cylindrical base as a more or less flat stone lying 'on the earth' and carrying the column shaft—which in the earliest temples was made of a wooden beam. This flat stone belongs more to the earth than to the space-creating and space-construction elements of the building. The simple cylindrical base is a most concentrated architectonical form, becoming its force by 'mother earth' on which it lies.[57]

The objections to the identification of Anaximander's "earth" with a column base are two fold: (1) the architect's module was column diameter and, as the archaeologists' debates show, that diameter is measured on the lower part of the column *above* the column base; and (2) Anaximander's idea of a free-floating earth, held up by nothing, is positively subverted by identifying the model with a base that, rather than "floating," is anchored to the earth. Let's examine these arguments in greater detail.

We know from Vitruvius that column diameter was the Ionic architect's module. This means that the architects selected one measured element in terms of which the rest of the architectural elements were reckoned as multiples or submultiples of it. This was the architect's One over Many; and since Anaximander measured his cosmos in column-drum proportions, he cosmically made use of the architect's temple module. Thus, the dimensions of the stylobate or the height of a column, for example, were each reckoned in column-drum proportions, as were the upper orders. But where precisely on the column is the modular measurement to be found? Since the column displays *entasis* as it is fashioned upwards, the diameter is not uniform throughout the column. Where, then, is the module to be located?

The debate among archaeologists centered on the precise location of the lower diameter where the module was established (fig. 2.8). Wesenberg argued that the module was located slightly above the usual identification of "ud" (*unterer Säulendurchmesser, gemessen oberhalb des Ablaufs*) and he identified it as the diameter at "UD" (*unterer Säulendurchmesser, gemessen auf dem Ablaufs*).[58] With this measurement as the module, Wesenberg set out to prove that the other reconstructed architectural elements now stood in the appropriate proportional relationships.

If the column is now seen in fuller form, it is clear that whether the module is located at "ud" or "UD," the location of the module is on the lowest column drum (fig. 2.9). But, at all events, it is not the base. Thus, the first argument against the hypothesis that Anaximander likely had in mind the metrologically determined and uniformly installed column base is that it was not the module that architects had selected.

The second argument for the identification of Anaximander's imagination with a column drum might begin this way. The most original part of Anaximander's cosmos is that the earth, according to Aristotle (*De Caelo* 295b 10ff), remains at rest in the center of the cosmos, held up by nothing. In view of this conception, identifying Anaximander's earth with the stone that sits on the ground betrays the very essence of his conception of an earth

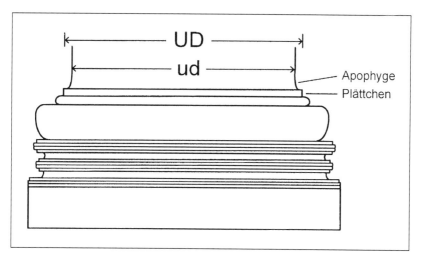

FIGURE 2.8. Where exactly on the lower column diameter was the module measured?

held aloft. The column base suggests more than any other part of the column the idea of something substantial that is nevertheless held up by some great support from below. And the technical problem of setting a column base is far more straightforward than installing a column drum. The base sits on the stylobate and is in danger of falling nowhere. It is held in place because of the substantial foundation below it, and to which it owes its support.

The column drum, however, stays in place by virtue of a clever technical solution analogous to the architect's *anathyrôsis* and *empolion*. Aristotle reports that Anaximander's earth is *homoiotêta*[59]—it is in equilibrium and is equidistant from all extremes and so has no reason to move up or down or to the sides.[60] By analogy, Anaximander's conception seems to draw upon just this technical problem that must have initially haunted the architects. The archaeological evidence shows us the architect's solution of how to solve the problem for each drum to remain in place, moving neither left nor right, neither up nor down. The technique required, as the drum face illustrates, a series of concentric circles by which the architect could exactly determine the center of the drum and the circumference band exactly *equidistant* from the center. Only by guaranteeing that condition of *homoiotês* could the architect solve the problem of each drum staying exactly in place, motionless.

Moreover, the Milesian *phusiologoi* sought to reveal nature's "hidden" structure.[61] In Anaximander's cosmic picture, he dared for the first time in the world of the ancient Greeks to claim that the earth, supported by nothing, remains motionless in place. Anaximander appealed to monumental temple architecture to reveal a structure that everyone in his audience could

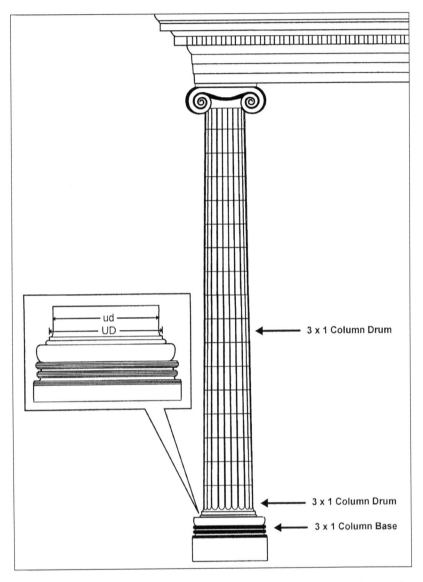

FIGURE 2.9. Where on the column did Anaximander imagine the column-drum earth?

be presumed to know, to have seen, and whose process of production was accessible to anyone who took the time to visit the building sites; then, by analogy, he sought to reveal nature's structure, which was nevertheless hidden knowingly in front of everyone's eyes.

Just as the Ionian architects must have astounded their communities by their techniques, and celebrated their achievements in prose books, so also Anaximander rivaled their excellence with his own prose book. While the books of the temple architects sought to illuminate how to build the house of the cosmic powers, Anaximander's book sought to explain the stages by which the great house that *is* the cosmos was built, the stages of construction by which it came to be.[62]

The method we are following here, then, is a kind of "inverse archaeology." Instead of starting with the archaeologists' artifacts and then trying to infer the abstract ideas that might have given rise to them, we begin with Anaximander's belief system and only then proceed to the archaeological evidence to see if clarification can be gained about the details in the doxographical reports. Given the reports that Anaximander identified the shape and size of the earth with a column drum, we appeal to the archaeologists' artifacts and reports to consider what Anaximander (and his community) would have observed when he called upon these products and the technologies that produced them. An appeal to archaeological evidence proves useful in supplying both a rich context and clues for how he brought together and developed his abstract speculations by means of metaphorical and analogical reasoning.

THREE

Anaximander's Cosmic Picture

The Homoios Earth, 9, and the Cosmic Wheels

A

THE DOXOGRAPHICAL REPORTS

Problem 1: The Earth is ὁμοίως in the Center

εἰσὶ δέ τινες οἳ διὰ τὴν ὁμοιότητά φασιν αὐτὴν (sc. τὴν γῆν)
μένειν, ὥσπερ τῶν ἀρχαίων Ἀναξίμανδρος. μᾶλλον μὲν γὰρ
οὐθὲν ἄνω ἢ εἰς τὰ πλάγια φέρεσθαι προσήκει τὸ ἐπὶ τοῦ μέσου
ἱδρυμένον καὶ ὁμοίως πρὸς τὰ ἔσχατα ἔχον. ἅμα δ᾽ ἀδύνατον εἰς
τἀναντία ποιεῖσθαι τὴν κίνησιν, ὥστ᾽ ἐξ ἀνάγκης μένειν.
—Aristotle, *De Caelo*, B13, 295b10ff, D-K 12A26.

*There are some who say that the earth remains in place because it is in
equilibrium, like Anaximander among the ancients. For a thing established in the center, and equally related to the extremes, has no reason
to move up or down or to the sides; and since it is impossible for it to
move simultaneously in opposite directions, it will necessarily remain
where it is.*

τὴν δὲ γῆν εἶναι μετέωρον ὑπὸ μηδενὸς κρατουμένην, μένουσαν δὲ διὰ τὴν ὁμοίαν πάντων ἀπόστασιν.
—Hippolytus, I,6.3; D-K 12A11.3

*The earth is aloft, held up by nothing; it remains in place because of its
similar [i.e., uniform] distance from all things.*

λίθῳ κίονι τὴν γῆν προσφερῆ.

—Aetius III, 10.2 [D-K 12A25, and B5]

The earth resembles a stone column.

Ἀναξιμάνδρῳ δὲ ἐδόκει καὶ διὰ τὸν ἀέρα τὸν ἀνέχοντα μένειν ἡ γῆ καὶ διὰ τὴν ἰσορροπίαν καὶ ὁμοιότητα.

—Simplicius, De Caelo, 532.14

Anaximander believed that the earth remains aloft because of the air supporting it, and also because of its equilibrium and uniformity.

Problem 2: The Cosmic Numbers

εἶναι δὲ τὸν κύκλον τοῦ ἡλίου ἑπτακαιεικοσαπλασίονα <τῆς γῆς>, ὀκτωκαιδεκαπλασίονα δὲ τὸν> τῆς σελήνης, καὶ ἀνωτάτω μὲν εἶναι τὸν ἥλιον, κατωτάτω δὲ τοὺς τῶν ἀπλανῶν ἀστέρων κύκλους.

—Hippolytus, I.6.5; D-K 12A11.5

The circle of the sun is 27 times the size of [the earth], that of the moon [18 times]; the sun is highest, and the circles of the fixed stars are lowest.

Ἀναξίμανδρος (sc. τὸν ἥλιόν φησι) κύκλον εἶναι ὀκτωκαιεικοσαπλασίονα τῆς γῆς.

—Aetius, II, 20.I; D-K 12A21

Anaximander (says the sun) is a circle 28 times the size of the earth.

Ἀναξίμανδρος (sc. φησι) τὸν μὲν ἥλιον ἴσον εἶναι τῇ γῇ, τὸν δὲ κύκλον ἀφ᾽ οὗ τὴν ἐκπνοὴν ἔχει καὶ ὑφ᾽ οὗ περιφέρεται ἑπτακαιεικοσαπλασίω τῆς γῆς.

—Aetius II, 21.I; D-K 12A21

Anaximander says the sun is equal to the earth, but that the circle from which it has its breathing-hole and by which it is carried is 27 times the size of the earth.

κύκλος εἶναι (sc. τῆς σελήνης) ἐννεακαιδεκαπλασίονα τῆς γῆς.

—Aetius II, 25.I; D-K 12A22

The circle of the moon is 19 times the size of the earth.

B

THE SCHOLARLY DEBATES OVER
THE TEXT AND ITS INTERPRETATION

THERE ARE TWO distinct problems that might best be resolved together: (1) *How* is it that a flat and cylindrical earth is in the center of the cosmos yet somehow "equidistant from" or in "equilibrium with" the cosmic extremes? (2) *What* exactly are the sizes of or distances to the heavenly wheels of stars, moon, and sun?[1] The underlying problem is to determine *how* Anaximander imagined the cosmic picture. What *model* of picturing underlies his cosmic imagination? From what sources of influence did he come to imagine the shape and size of the cosmos?

The first problem arises from Aristotle's testimony in *De Caelo* that Anaximander, distinguished among the ancients, maintained that the earth is at the center of the cosmos, *held up by nothing*. It remains there because there is *no reason* for it to move either left or right, up or down. Thus, the earth for Anaximander remains in equipoise—*homoiôs pros ta eschata echon*[2]—equidistant from extremes. Commentators have been troubled by Aristotle's testimony for a variety of reasons; a persistent and underlying problem has been to account for the apparent geometrical incongruity of maintaining that Anaximander's earth is equidistant from the extremes, which might be plausible if the earth were spherical, while acknowledging also that Anaximander's earth is both flat and cylindrical. This problem, among others, as some scholars have argued, seems to lead to one of two defeating conclusions: either Aristotle is misreporting Anaximander, or Aristotle is reporting correctly but Anaximander's own geometrical imagination is flawed. Is there yet a way to salvage Aristotle's testimony and save Anaximander's conception of the earth as *equably related to the extremes*?

The second problem concerns the reports about the sizes of or distances to the stars, moon, and sun. Anaximander's cosmic picture identifies the stars, moon, and sun as fiery wheels encased in and thus concealed by mist or *aer*. Scholars usually posit that the numbers attached to the wheels are 9, 18, 27, and so present a geometrical progression from the 3:1 earth of increasing distances and sizes. But the doxographical reports we actually have provide no numbers at all to describe the size of or distance to the stars, though we have the testimony that they are closer to us than the moon or sun.[3] The number 9, then, is an interpolation offered to complete a series based in the number 3. But what exactly is the series? The numbers 27 and 28 are *both* assigned to the largest and most distant wheel, that of the sun. The number 19, not 18, is assigned to the moon wheel. Can we resolve the problem of the cosmic numbers?

Thus, we have two problems: (1) to resolve the apparent geometrical conundrum of explaining how a flat and cylindrical earth, unlike a sphere, could be equidistant from all extremes, and (2) to determine precisely the cosmic numbers for the heavenly wheels that describe the sizes of or distances to them.

What has been missing from the scholarly literature is a serious reflection on *how* a Greek of the archaic period could have imagined the cosmos. What evidence do we possess that might suggest *how* Anaximander imagined, what techniques of modeling informed his cosmic imagination? When we consult the archaeological record, a resource that scholars of ancient philosophy routinely ignore, we see an approach that offers a clue for resolving both difficulties at once. We shall see, more clearly yet, new background questions that need to be answered and, until recently, had never before been raised in the scholarly debate.[4] When we try to imagine Anaximander's cosmic picture, from what point of view is he picturing? When Anaximander describes the earth, stars, moon, and sun, where is he standing? Is he "outside" the cosmos looking in? And, if so, from where? From above? From below? From a side? Is Anaximander imagining while standing on the earth and looking up, or down? What evidence is there to show *how* Anaximander was imagining? To engage this question, we need to determine what point or points of view would have been meaningful to an archaic Greek in general, and to Anaximander in particular. What evidence do we have that could be telling?

B.1

The Earth is Homoiôs (ὁμοίως) in the Center

According to Aristotle in the *De Caelo* passage, Anaximander held that the earth is stationary in the middle and that it is "equably related to the extremes." Because it is equally related to the extremes, according to Aristotle's report, Anaximander's earth should not be inclined to move upwards or downwards, or sideways—that is, left or right. This suggests that in order for Anaximander's earth to move one way or the other, there needs to be a *reason* for it to do so; situated in the center, however, there is no reason for it to move in one direction or another. Thus, we have what is regarded by some scholars to be the earliest surviving example of the Principle of Sufficient Reason.[5] Besides Aristotle, we have the testimony of Hippolytus that Anaximander's earth stays in the center *dia tên homoian pantôn apostasin*,[6] by reason of its equidistance from all things, and also from Diogenes Laertius that *mesên te tên gên keisthai, kentrou taxin epechousan*[7] and the Suda: *Anaximandros Praxiadou Milesiou prôtos de . . . heure . . . tên gên en mesaitatô keisthai*.[8] Since Hippolytus, and this later tradition, presumably relies on Theophrastus, and hence ultimately Aristotle, objections to Aristotle's claims hold the promise of undermining Aristotle's remarkable account itself.

The scholarly debate provides a panoply of issues. The argument attributed to Anaximander is so remarkable at this early date as to be considered unbelievable by some. Is Aristotle misreporting, thinking instead of the passages from Plato's *Timaeus* and *Phaedo* that he had been discussing just prior to referring to Anaximander in *De Caelo*? Passages in both Platonic dialogues mention the earth's immobility and, moreover, that a thing in equilibrium—*homoiôs d' echon*[9]—and occupying a place at the center, will not be inclined to either side (*aklines*).[10] But, in those Platonic discussions, it is a spherical earth that is in equilibrium, not Anaximander's flat and cylindrical earth. But might it be that Aristotle is misreporting in another way? Perhaps Anaximander's earth is in the center, but not by the argument Aristotle attributes to him. It might be the vortex, the *dinê*,[11] that accounts for its being in the center, as it functions for other early Greek thinkers. And even if Anaximander's earth is in the center, it still remains for us to get clear about the "center" of what. Robinson argues that if Anaximander's earth is *equably related to the extremes*, it must be in relation to the circumference of the *ouranos*.[12] Furley argues, to the contrary, that *homoiôs* can only mean in relation to the heavenly wheels.[13] Aristotle attributes to Anaximander what seems a most remarkable doctrine. How do we resolve these issues?

After reviewing the wealth of scholarly interpretations, there seem to be three issues at stake that suggest three competing theories to explain how Anaximander's earth is in the center: (1) Does the earth come to be in the center because of the *dinê* ? (2) Is the earth held aloft because it floats on air? (3) Is the earth at the center because it is *homoiôs pros ta eschata echon*?[14]

Heidel doubts Aristotle's report and favored the *dinê*, but he also recognizes a problem with this view.[15] Had the *dinê* forced the cold and moist earth to the center, we should expect the heavenly wheels to be in the same plane as the earth, which they clearly are not; had they all been in the same plane, the sun, moon, and stars would never rise above the horizon and so would never be visible in the way that they in fact are. Consequently, Heidel tried to salvage the *dinê* theory by positing an *ekklisis*,[16] that is, a dip of the earth to the south, to explain, without pain of inconsistency, how the sun and moon could appear above the horizon.[17] Thus, Heidel's solution requires us to imagine that the earth and heavenly wheels are in fact all in the same plane but do not appear that way because of the *ekklisis*. This view was defended in a later edition of Burnet[18] and also by Rescher.[19] The most formidable problem with this approach is that there is no direct evidence that Anaximander held this position on the *dinê*,[20] though it is clear that for him "motion is eternal."[21]

Robinson also doubted Aristotle's testimony, and relying on Simplicius' claim that the earth was in equipoise <u>and</u> *supported by air below*, argued that Aristotle failed to distinguish two separate problems: (1) the problem of lateral fixity—why the earth does not move left or right; and (2) the problem of vertical fixity—why the earth does not move up or down. Robinson reaches the conclusion that Aristotle failed to grasp that the problem of vertical fixity required that Anaximander held that the earth had absolute weight and would have fallen down had it not floated on air. The air below supported the heavy earth in a manner akin to Anaximenes' solution.[22] Robinson's position on vertical fixity has also been defended by Furley. Furley appeals to Simplicius' testimony, and argues that the Milesians were impressed by the ability of objects to remain afloat because they were flat; he takes the issue of "equipoise" to mean only that the earth was not inclined to tilt one way or the other but remains aloft (i.e., stable and level) by the support of the air under it. But this position has its problems as well. The only testimony that Anaximander's earth remains aloft because of the air that supports it is Simplicius, who in the same breath adds to the reasons why the earth hangs freely in space "because of its equilibrium and uniformity" (*dia tên isorrhopian kai homoiotêta*).[23]

These two competing theories, perhaps, are *not* mutually exclusive. The appeal to the *dinê* offers to explain the *lateral fixity*, how the earth is located at the center; the appeal to the support of the air from underneath offers to explain the earth's *vertical fixity*. What is clear, however, is that the appeal to either the *dinê* or the "support of air from below" undermines Aristotle's *homoiôs* account. For both alternatives appeal to some sort of "force" that holds the earth in place at the center. Hippolytus reports that *tên de gên einai meteôron hypo mêdenos kratoumenên, menousan de dia tên homoian pantôn apostasin*,[24] that is, it is not dominated by anything, it is not held in place by force. If we accept Hippolytus' testimony along with Aristotle's, these two alternatives must be dismissed to explain the earth's immobility at the center. This does not require, however, that we dismiss the attribution of the *dinê* to Anaximander, as we must discredit Simplicius' report about the supporting air underneath; it only requires that we dismiss the *dinê* as an explanation for *how* the earth comes to remain at the center.[25]

The defense of Aristotle's testimony, along with Hippolytus', has been robust. Kirk-Raven-Schofield,[26] Kahn,[27] Guthrie,[28] Vernant,[29] West,[30] Lloyd,[31] McKirahan,[32] Naddaf,[33] and Couprie,[34] among many others, have defended the view. While it is true that the other doxographical reports that lend support to Aristotle's testimony might ultimately derive from Aristotle—such as Hippolytus, but also Diogenes Laertius,[35] the Suda,[36] and Theon of Smyrna[37]—only the one isolated testimony by Simplicius adds to the equilibrium theory the appeal to the support of air below. Kahn offers the clearest defense of Aristotle's report and against the support from the "air below" thesis: The Simplicius passage occurs in a context that does not suggest a direct consultation of Theophrastus. Simplicius' assertion has no parallel in other doxographical reports, and is implicitly denied by Aristotle's omission of Anaximander in the list of those who do make use of air underneath for support at *De Caelo* 294b13,[38] the passage just before Anaximander is distinguished from the others.

The defense of the *homoiôs* reading has yet another appeal, namely, the testimony about the so-called antipodes. Hippolytus reports that *tôn de epipedôn hô men epebebêkamen, ho de antitheton hyparchei.*[39] Referring to Anaximander's earth, Hippolytus says that on the flat side we walk, and opposite there is another side. Whether or not Anaximander held that the opposite side of the earth was inhabited, what is clear is that his position undermines any absolute sense of up or down, left or right, and thus suggests a new vision of "space" in which the earth can be *equably related to the extremes.* Anaximander's new vision of space, original in the context of archaic Greece, offers a mathematical sense—or perhaps we should say a geometrical sense—of space, a radical departure from the mythic worldview.

In a thoughtful essay, Vernant[40] emphasizes the originality of Anaximander's conception of space. He observes that mythical space is expressed in the following tri-fold hierarchy.

THREE-TIERED MYTHIC SPACE

(1) The heavens ABOVE the earth:
Zeus and the Olympian gods

(2) On the earth:
Man (*en meso*)

(3) BELOW the earth:
the dead and subterranean gods

Zeus and the Olympian gods are highest, man resides in the middle, and the dead and subterranean gods remain below. The world so conceived does not readily allow, except in special circumstances, the movement from one level to another. In contradistinction, Anaximander's originality is announced, first of all, by his claim that there is "space" itself; this is unlike the earlier view in Homer of a celestial dome with all heavenly objects uniformly distant.[41] Anaximander posits that the stars are very far from us, and *behind* the stars lies the moon, and *behind* the moon lies the sun farther still. In Hesiod's mythical space, the region below the earth was likened to a huge jar at whose mouth grew up the roots of the world.[42] In Xenophanes' visualization of the space below the earth, we learn that it extends downwards forever.[43] In the midst of these views, Anaximander dared to maintain that the fiery wheels of sun, moon, and stars go *underneath* the earth, and that the earth remained aloft in the center, held up by nothing. This is an astonishing and new view that Anaximander advanced. *How* did he reach this conclusion?

Vernant offers to explain Anaximander's new vision in the context of social and economic transformations that were broadly contemporaneous; Anaximander imported into his reflections about nature, about *phusis*, the newly developing reflections about social life.[44] The development of the polis, according to Vernant, proved to be a process of the secularization and rationalization of social life. Greek cosmology was able to free itself from religion because knowledge concerning both nature and social life became secularized and rationalized at the same time. This leads Vernant into a detailed reflection about the efforts in city planning to appropriately divide the regions of social life; the trifold division of the heavens into sun, moon, and stars mirrored the rationalizing city planning of a trifold division of the city. Naddaf echoes Vernant's approach and suggests that Anaximander's cosmic layout was reasoned by analogy with the layout of the proposed, idealized city.[45] Both Vernant and Naddaf, then, envision Anaximander to have politicized the cosmos with the earth *en meso*.

B.2

THE COSMIC NUMBERS

Simplicius reports, not on the authority of Theophrastus but rather Eudemus' ἀστρολογικὴ ἱστορία, that Anaximander was the first to describe the sizes and distances (*peri megethôn kai apostêmatôn*)[46] of the heavenly bodies.[47] According to Hippolytus, in a text that seems almost certain to contain some corruption,[48] the circle of the sun is 27 times the size of the earth and that of the moon 19 times, and the sun is the highest of the heavenly bodies while the stars are the lowest. Aetius says, first, that the circle of the sun is 28 times the size of the earth, and then shortly thereafter that the sun is equal to the earth, but the circle from which it has its "breathing hole" is 27 times as great as the earth. Aetius also reports in the same passage the order of distances, noting that the sun is highest, next the moon, and beneath them the fixed stars. What are we to make of these reports?

Tannery,[49] Diels,[50] Burnet, and Heath agreed that the numbers provided in the doxographical reports could all be accepted. They shared the conclusion that the number 27 refers to the inner circumference of the sun's circle while 28 refers to the outer circumference. The breadth of each heavenly wheel is reckoned as one earth-diameter, since the "sun is equal to the earth." This would mean, consequently, that the number 19 refers to the outer circumference of the moon's circle and that the number 18 could be justifiably supplied to the inner circumference. And since the stars are closer to us than either the sun or moon, the rest of the series is inferred as 9 to the inner circumference, and 10 to the outer circumference. There have been two different kinds of reasons for defending this reading. On the one hand, Burnet suggested that the numbers 9, 18, 27 "play a considerable role in primitive cosmogonies";[51] in much the same vein, Heath emphasized also that the number 3 and the series 9, 18, 27 have symbolic meaning: "These figures suggest that they were not arrived at by any calculation based on geometrical constructions, but that we have merely an illustration of the ancient cults of the sacred numbers 3 and 9."[52] A detailed exegesis has also been made recently tracing the significance of the number 3 in ancient cosmogonies, and this collection of evidence tends to bolster the case.[53] On the other hand, there are those like Kahn, Burkert, and McKirahan, who regard the series as expressing Anaximander's mathematical and geometrical mentality, rather than emphasizing the symbolic meaning that could be traced to admittedly primitive cosmogonies. Kahn claims that "Anaximander clearly believed the universe was governed by mathematical ratios";[54] Burkert asserts that "the world of Anaximander . . . is essentially geometrical;"[55] and McKirahan maintains that "he [Anaximander] assumes that the KOSMOS has a simple geometrical structure."[56] It would be fair to say, then, that the chorus of commentators accepts that the heavenly wheels stand in increasing distances from the 3 × 1 column-drum earth—stars, moon, and sun—and that the numbers 9, 18, 27 are appropriately applied to them. And it would be fair to add that the chorus agrees also with the sentiment stated most stridently by Dicks, that the numbers cannot have been based on any accurate astronomical observations.[57] The one caveat that needs to be made, however, is this. While Anaximander's cosmic numbers were not based on what we now regard to be astronomical observations, the case being argued here is that the results could not confound commonsense observations (a point whose implications will be considered shortly) and, as the reconstruction of his seasonal sundial will show, Anaximander was capable of making careful observations that could be tested empirically, and by everyone in the community.[58]

Finally, we might add one other consideration about the ordering of the heavenly wheels, though it will not explicitly affect the cosmic numbers. There are two diverging theories to account for the ordering that places the stars closer to us than the moon or sun, an ordering that has no clear antecedent in Greek meteorology. There is the "rational" interpretation advanced by Kahn and others that the ordering reflects the nature of "fire," that is, heat naturally rises or goes up. Thus, where there is more fire, the heavenly wheel is higher (i.e., farther); the sun is hotter, and thus farther while the moon and star wheels have less fire and are thus closer.[59] On the other hand, there is the "Iranian/Persian" interpretation of the *Zend-Avesta*. Burkert argued that there was a parallel set of passages, though of earlier date, that explain how Ahura Mazda travels to the sun, then the moon, then the stars, and finally to your *hestia*, the hearth in your home.[60] West defends the view that there was Persian importation into Anaximander's thought;

Anaximander had integrated into a tradition of Greek meteorology a tradition of Zoroastrian cosmic speculation. As West elegantly stated the matter:
"Anaximander's conceptions cannot be derived from Greek antecedents, and
to suppose that they chanced to burgeon in his mind without antecedents, at
the very moment when the Persians were knocking at Ionian doors, would be
as preposterous as it was pointless."[61] Thus, the rival interpretation to the
"rational" ordering of the heavenly, fiery wheels is to credit the sequence of
sun, moon, and stars to Anaximander's importation of Zoroastrian cosmological speculation. But, in objection to this rival interpretation, at least one
scholar suggests that the Avestan passages to which these adherents refer is
later than Anaximander, and so it has also been conjectured that perhaps it
is Anaximander who influenced the Avestan tradition and not vice versa.[62]

Now concerning the numbers, not all those who accept some series of 9,
18, 27 accept the arguments for 9/10, 18/19, 27/28 propounded by Tannery,
Diels, Burnet, and Heath. Kahn, for example, regards the occurrence of 28 in
the first passage by Aetius as a corruption. He acknowledges that others have
defended this reading, and believes the chief advantage of this interpretation
is that "it would permit us to replace the awkward figure 19 for the circle of
the moon by 18 (performing the same subtraction in order to find the 'inner'
diameter) . . . giv[ing] us the neat series 27:18:9, whose last term should represent the diameter of the stellar rings."[63] But Kahn rejects this interpretation
on the grounds that it lacks documentary basis.

West proposes the most ingenious objection to the conventional reading, agreeing with Kahn that the number 28 is a corruption but also that 19
is a corruption as well.[64] First he points out that while Aetius discusses the
sun and moon in separate sections, he suggests that the original source was
unlikely recorded in this way but rather was presented together in one passage. Thus, he proposes a reconstruction of the original: "Suppose that the
original version ran εἶναι δὲ τὸν τοῦ ἡλίου κύκλον ἑπτακαιεικοσαπλασίονα τῆς γῆς, τὸν δὲ τῆς σελήνης ὀκτωκαιδεκεπλασίονα, τοὺς δὲ τῶν
ἄστρων ἐννεαπλασίονας. Suppose that a scribe wrote ἐννεα– instead of ὀκτωκαιδεκεπλασίονα, his eye running on to ἐννεαπλασίονας. ὀκτω was then
written in the margin as a correction, but mistakenly substituted for ἑπτα in
ἑπτακαιεικοσαπλασίονα."[65]

As we proceed to a proposed resolution, we need to ask whether the
term *plasiona* (πλασίονα) refers to the earth's diameter, circumference, or
radius. And what difference does it make? Diels,[66] Dreyer,[67] Mieli,[68] Burch,[69]
Kahn,[70] and Guthrie[71] construed *plasiona* as diameter. Zeller[72] refers to circumferences, and so does Burnet[73] in his fourth edition, but in his first edition he refers to diameters. Taylor[74] speaks of radii, as does Tannery in one
work, but in another work Tannery refers to diameters,[75] and most recently
Couprie also maintained that the numbers must be radii.[76] What is at stake
is this: If *plasiona* means diameter or circumference, then Anaximander was

referring to *sizes* of the heavenly wheels; if *plasiona* means radius, Anaximander was referring to *distances* to them.

Kirk recognized that there was a problem with the doxographical reports concerning the cosmic numbers. If we are to accept both 28 and 27 to identify the sun wheel, they cannot refer to "diameters" because even accepting the width of the wheel as one earth-diameter, the outside number should have been 29. Then Kirk reaches a precipice: "If the radius and not the diameter were intended the figures would hold" but he concludes that it seems that diameters are meant. This was Kirk's mistake. Anaximander stands on the edge of a new tradition in Greek cosmology and astronomy—the discovery of space, that there is depth in space. Seen in this way, he was concerned first and foremost with identifying the *distances to*, not the sizes of, the heavenly wheels. The sizes of the wheels have implications quite distinct, and a case has already been made for their analogous and metrological relation to the temple *kernbau* and interior structures—that is, the relation of the inner kernel of the temple to its outer structures.[77] As we will explore in the appeal to the archaeologist, the architectural technique favors a construal that the numbers are radii, not diameters, once we get clear about the model that allowed Anaximander to imagine the cosmos. And the appeal to what archaeologists term "technological style" will further support the interpretation of the cosmic numbers to give distances based on the archaic formula of 9 + 1 and 9.

There is one other consideration—the observed angular diameter of the sun—that bears directly on the cosmic numbers and Anaximander's "astronomical observations." Diogenes Laertius reports that, according to some authorities, Thales was the first to declare the apparent size of the sun to be 1/720th part of the circle described by it.[78] On the surface, this is hard to understand because we have no indication that Thales believed that the semicircle that the sun traversed continued under the earth and so must be at best an inference from a semicircular projection of a circle that it might describe. And it is strange also that no one before Archimedes mentions this correct result,[79] and so some scholars have conjectured that Thales, had he actually made this proposal, was relating information from Egyptian or Babylonian sources. But from the secure evidence we do have, Aristarchus had discovered, some three centuries *after* Thales and Anaximander, the value of 1/720th of a circle, or 1/2°, for the angular diameter of the sun; however, in his treatise, Aristarchus offered the value of 2° for the angular diameter of the sun for other calculations.[80] Now, if we regard Anaximander's cosmic numbers to be radii and not diameters, the numbers would allow us to measure the angular diameter of the sun at 2°. Let us consider what this means.

While Anaximander's cosmic numbers are not based on observational techniques, appealing instead to a poetic formula, the results could not confound common observations. Heath pointed out that if the cosmic numbers were interpreted as "diameters," and the sun was 28 (or 27) times the size of the earth, while the sun is the "same size as the earth," the sun's apparent diameter would be a fraction of the whole circumference of the circular wheel expressed as 1/28, and thus the angular diameter would be 360°/88 or a little over 4°, which is eight times too large and which Heath himself acknowledges would be rejected by common observation.[81] Zeller sought to repair the grossly excessive result by wondering if the sun's circle should be 27 times the moon's circle, making it 513 times the size of the earth; not only is there no textual support for this conjecture but, moreover, the result would make the angular diameter of the sun much too small. But if we suppose that the doxographical reports are garbled, and *plasiona* refers to radii, not diameters or circumference, then the angular diameter of the sun would be $1/2\pi r$ (= $1/2 \times \pi \times 27$), or 360°/162 or slightly over 2°, which, while still making it some four times too large, is still in concert with what Aristarchus regarded as suitable for cosmic calculations almost three centuries later. This observational result, though incorrect, is not an excessive exaggeration for common observations. Interpreted within this context, the reading that the cosmic numbers are radii is supported further.

C

THE ARCHAEOLOGICAL EVIDENCE

There are two archaeological resources that offer the prospect of clarifying, and perhaps resolving, the scholarly debates on these problems, but the plausibility of both requires that we take as our starting point the supposition that Anaximander imagined the cosmos as "cosmic architecture" and so imagined it by means of architectural techniques he could have witnessed at the building sites: (1) the architectural technique for laying out the stylobate of an archaic temple, that is, a *plan-view* approach to column diameter measurements, and (2) the appeal to what the archaeologists call "technological style" to confirm the numbers for the heavenly wheels. At all events, the point of departure is to acknowledge that Anaximander was present at the building sites, watched the architects and builders at work, and applied these techniques to his cosmic reflections *because he imagined the problem by analogy to be cosmic architecture*.

C.1

ARCHITECTURAL PLAN TECHNIQUES
FOR LAYING OUT THE GROUND PLAN

The case has already been made that Anaximander was drawing on firsthand knowledge from architectural building sites, most likely at the temple of Apollo at Didyma and the cult buildings along the Sacred Road connecting the sacred precinct to the mother city of Miletos. But, considering that various doxographical accounts unmistakably suggest that Anaximander was a much-traveled man, it is no less believable that he visited the nearby great temple building sites in Ephesos and Samos; that is, it is not unlikely that he had firsthand knowledge of the *general* process of the planning and execution of the constructions.[82] Whether he was a more active member of the architectural and building teams is a more difficult case to make, but it is possible, especially when we consider that the tertiary evidence we do have about Anaximander's interests and activities shows he was engaged in a range of problem solving, often requiring applications of practical geometrical skills. Since so much of ancient architecture entails techniques we might speak of broadly as "applied geometry," it would not be surprising if Anaximander took a keen interest in architectural problem solving. What is clear at the very least is this: Anaximander was present at building sites and he was paying careful attention.

The case for this far-reaching discovery begins with the simple clue that Anaximander had identified the shape and size of the earth with a column-drum at the very same time when column drum construction was introduced to Ionia. Vitruvius reports that the module—the One over Many—for the Ionian architects was "column diameter," and this means that other dimensions of the building were reckoned in column-drum proportions.[83] Moreover, we learn from Vitruvius that the archaic architects of the sixth century BCE wrote prose books.[84] Now, Anaximander measured the cosmos in earthly, that is, column-drum proportions, and thus he not only adopted a modular technique for expressing cosmic dimensions but also precisely adopted the architects' own module. The import of this coincidence is impossible to overestimate. That Anaximander wrote a prose book more or less contemporaneously with the architects suggests also that he was part of their "community" or they of his, in a way that has never before been appreciated. The coincidence of their publication of prose books raises the question about who wrote first—that is, who might have inspired the other to make public their prose writings. It is tempting to suppose that the architects first made public their writings about the temple, the proportions and stages by which the house of the cosmic powers was built, and that Anaximander joined their enterprise, writing for an audience already interested in discussions about proportions and numbers, but this time about cosmic architecture—the stages by which the house that *is* the cosmos was built.[85]

Granting that Anaximander was present at the building sites and at least watched attentively, let us consider what techniques he would have observed that might illuminate the ongoing scholarly debates about the earth's position and cosmic numbers of the heavenly wheels. The background question is: *What model of imagining did Anaximander have in mind when he sought to imagine the cosmos?* The answer proposed here is that Anaximander focused upon the architect's technique of "plan-view" construction and the modular technique of setting out the temple ground plan. Then, he adopted and adapted these techniques for analogously imagining the layout of the cosmic ground-plan.

However, the temple "ground plan" and the cosmic "ground plan," so to speak, were also disanalogous in important ways: While the temple has an absolute ground, and hence is amenable to absolute directions of above and below, right and left, Anaximander's cosmos has no absolute directions. The earth remains in the center, held up by nothing. It is not implausible that Anaximander held that the antipodes are on the opposite side of the earth on which we stand. In this case, for the antipodes there is an open space "above" not below, just as there is for us. Consequently, to speak of a cosmic "ground plan" would be misleading, unless we refer, instead, to a *horizontal cross section* through the plane of the earth as a focal point for Anaximander's cosmic architecture, because there is no terra firma upon which the earth itself rests. What this means is that, as we proceed, we must realize that while Anaximander, and his audience, marveled at the ingenuity of the temple architects and their successful techniques for making their *thaumata*, he chose to adapt them while explaining the architecture of the cosmos—the greatest *thaumata* of all—and the stages by which the present world order had come about. While temple architecture is replete with right angles, squares, and straight lines, cosmic architecture is fundamentally cylindrical and circular. The planning and installation of the temple's colonnade, of course, offered images of circles and cylinders for the architect's reflection, and Anaximander who observed them. The result was that Anaximander undertook an enormous and bold imaginative project, supplying techniques from the architects but adapting them to an architecture—the cosmos—that displayed predominantly distinct geometrical forms.

The evidence for plan-view techniques in ancient architecture has already been set out.[86] In the absence of any truly monumental architecture in Greece, and certainly Ionia, for hundreds of years since the fall of Mycenae, the inspiration and techniques for monumental temple building were likely imported from Egypt. The Milesians had gained great repute for seafaring exploits and had supplied ships along with mercenaries to assist Pharoah Psamtik to regain the throne at his capital in Sais. In nearby Naucratis, Miletos, alone among the Greek colonies, was granted the right to establish a trading post. From there, the Milesians found the inspiration for multi-columned monumental temples and architectural guidance to build them, supplied, no doubt, by a grateful pharaoh. The evidence for "plan-view" techniques in architecture that survives from Egypt is impressive and compelling, thanks to that arid climate; it shows that plan-view imagining was both required and familiar to the ancient Egyptian builders over thousands of years prior to the arrival of the Milesians.

The surviving evidence for plan-view techniques in Greece, however, is exiguous. The case for plan-view techniques, nonetheless, can be made robustly. Because the archaeology of the ground plan is often the only surviving part of an archaic monument, we have a significant amount of evidence from which we can derive details about numbers and proportions on the stylobate. From these numbers and proportions that are derived directly from the excavated ground plan, the other dimensions of the building in elevation are plausibly conjectured. The excavated evidence convincingly points to the techniques by which the temples and cult buildings were produced. Working with informal sketches with numbers and measures, probably on papyrus or animal skins, the Greek architects set out the ground plan. Using measured rope or string, as well as stakes, the architects laid out the ground plan, 1:1. Then, they made scratch marks on the stylobate to indicate the placement of column bases and cella walls. What the archaeologists show us is that when careful measurements are made of the remaining ground plan, an informal sketch with numbers and measures can be inferred. Let us be very clear about this. Architects did not get up one morning and bring measured stakes and cords to the building site and start setting out their gigantic temple without a carefully thought-out plan—they had a sketch with numbers and measures to remind them, and almost certainly a model of the finished building probably made out of impermanent materials.[87] Figure 3.1 is an archaeologist's reconstruction of the sixth-century BCE ground plan of a cult building in Didyma on the Sacred Road, and below it the metrological reconstruction of the grid plan—that is, the informal sketch with numbers that is inferred.

The *plan* technique required that the architects *imagine*. The plan assumes a position 90° above the building, as if one were in a hot-air balloon. Since a plan requires a position that no person in antiquity could actually occupy, the technique reminds us of the architect's imagination. To control the construction, the architect must imagine each level of the building from the stylobate on upwards; Anaximander's cosmic architecture is a testimony not only to the same imaginative skills, but also the same model of imagining.

How did the temple architects imagine and set out the ground plan? The details have already been discussed at great length, though the accounts must remain, in part, conjectural. In the archaic period, where there is more likelihood that individuals or prominent families were patrons of the temples, rather than the public building projects that developed in classical times,[88] the patron either called out the total number of columns or the size of the stylobate. Then, a decision was made about a basic unit—a module—in terms of which other architectural elements would be reckoned. Vitruvius tells us that "column diameter" was the module for the Ionic temples. This would mean that once a decision was made about the column diameter,[89] then the

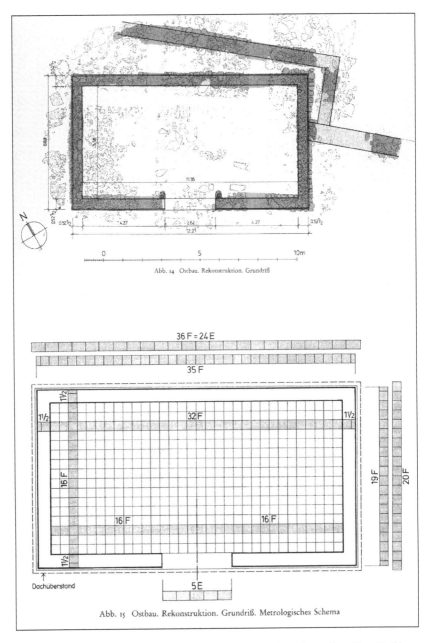

Abb. 14 Ostbau. Rekonstruktion. Grundriß

Abb. 15 Ostbau. Rekonstruktion. Grundriß. Metrologisches Schema

FIGURE 3.1. (top) Reconstruction of the ground plan from the archaic East Building on the Sacred Road connecting Miletos to Didyma; (bottom) informal grid drawing reconstructed from which the architect worked.

architect had to determine the spacing between the columns, and then, by successive addition or successive division, determine the size of the stylobate. Since the defining feature of temple architecture was the flow of dark and light spaces achieved by column width and intercolumniation, the spacing between the columns can be expressed by a multiple of column diameters. Thus, for aesthetic reasons, a specific intercolumn spacing was selected. When Anaximander watched architects work, laying out this part of the ground plan, what would he have seen? That is, how was the spacing between the columns—the intercolumniation—measured? What techniques would he have witnessed?

Archaeologists have suggested two approaches; following the lead of architectural historians, we can call one of them "intercolumnar" and the other "interaxial." If the architect measured the intercolumnar distance (fig. 3.2), this means he measured from outside edge to outside edge of each column. There are two ways to imagine this measurement: either the archaic architect of the great Ionic temples measured from lower column diameter to the next lower column diameter, or the measurement was made from the edge of the plinth to the next plinth, the square platform on top of which

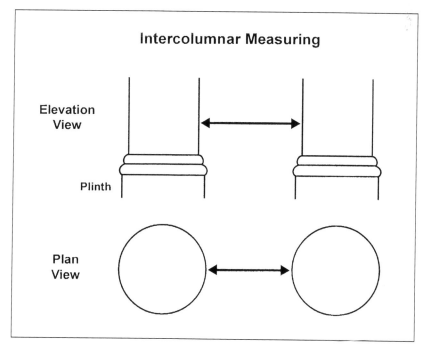

FIGURE 3.2. Plan and elevation views of intercolumnar measuring.

the column was installed. If, instead, the architect measured interaxially (fig. 3.3), he would have measured from the center of one column to the center of the one next to it.

The architect required knowledge of both measurements. Why? The architect needed to know the interaxial distance because that would determine precisely the length of the architrave blocks that would be installed above the columns, with the edge of each architrave block terminating at the interaxial center. In Doric architecture, for instance, there was another complication; the length of the architrave blocks is also tied to frieze units, and so another calculation was necessary to accommodate the placement of these architectural elements above the architrave. In Ionic architecture, the details are different, but the upper parts of the order depended also on the interaxial measurement. Thus, for these reasons, the architect would inscribe center to center on the stylobate after marking the column diameter.

But the architect also needed to know the intercolumnar distance for two different reasons: first, since the overall aesthetic of the temple was defined by visual impact of the column spacing, the intercolumnar distance would be pivotal to achieving the desired aesthetic effect of alternating opened and closed spaces; second, the architect was saddled with the real

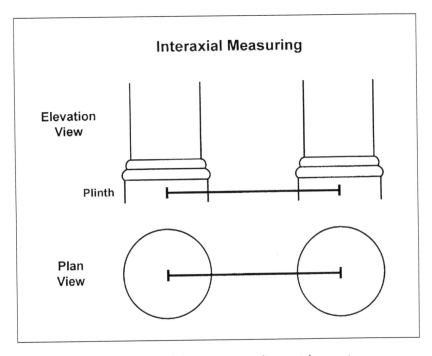

FIGURE 3.3. Plan and elevation views of interaxial measuring.

problem of spanning the architraves. In part, this means that the limits of architrave length were constrained by the material; thus, the marble used at the Ephesian Artemision could span a greater length than the limestone used at the Samian Heraion, while still being able to carry the load of the architectural elements above. But interaxial measurements are insufficient to resolve the problem of spanning lengths since it is not the center of the column, but rather the extending width of the architrave over the column extremities and the column capital that is crucial to this determination. Thus, the architect needed to know both the intercolumnar and interaxial distances to imagine and effectively execute the temple design.

Given these architectural measurements, the architect laid out the ground plan. By reference to a grid drawing, an informal sketch with the numbers he needed, he used a measured rope or string and set out stakes as he began to work.[90] Figure 3.4 is a reconstruction of the Hellenistic temple to Apollo at Didyma. Although this is the later temple, it is quite similar; moreover, von Gerkan's reconstruction provides the kind of grid drawing that the architects plausibly used and that Anaximander consequently would have seen. We must keep in mind that von Gerkan's reconstruction is the grid drawing inferred from the archaeologist's excavations.

The proportions of temples in *elevation* are crucial to what they looked like, of course, but the proportions of temples in ground plan are the ones

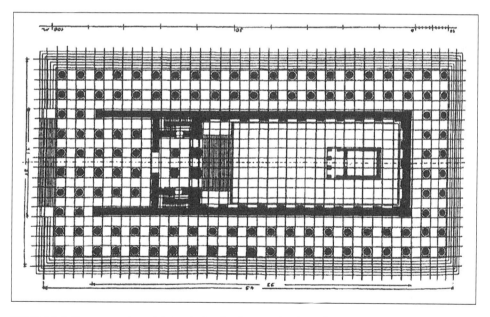

FIGURE 3.4. Reconstruction of the grid plan for the temple of Apollo at Didyma, after von Gerkan.

that are most studied because more often than not this is all that remains on many sites. But it was the slenderness and spacing of columns that most affected the appearance of the finished buildings and thus were most critically refined by the ancient architects. This brings us inescapably to column diameters and the spaces between the columns. If the architect specifies an intercolumnar or interaxial distance, he must also specify the (lower) drum diameter to establish any desired proportion. By *using drum diameter only, and a multiple of it for the space between the columns, one "unit" can establish the proportional system for a temple of any size.* This was the basic lesson and technique that Anaximander would have witnessed at all the temple building sites in Ionia, and it was this technique that he adapted when he sought to imagine cosmic architecture.

Starting with the column-drum–shaped earth in the center, Anaximander imagined the cosmos in plan view, that is, a horizontal cross section through the plane of the earth. Anaximander's earth, we are told, is *homoiôs pros ta eschata exon.*[91] The objection that only a spherical earth, not a flat cylinder, could fulfill this condition is vitiated once we see that Anaximander imagined the cosmos from more than one point of view. This is a most important point missed by all the commentators, and it was missed because commentators neither inquired about the techniques for modeling that burgeoned in Anaximander's mind nor explored the architect's techniques for modeling that were in use contemporaneously in Didyma, Ephesos, and Samos. There is no discussion in the scholarly literature about whether Anaximander imagined the cosmos from more than one point of view. Indeed, there is no discussion about what point or points of view would have been meaningful to a Greek of the archaic period. This unasked question and its resolution are central to this study. In Anaximander's imagining of the cosmic picture, he appealed to architectural techniques. Imagining a cross section through the plane of the earth—a plan-view technique—he could proceed as if the heavenly wheels lie obliquely to the surface of the earth, though in fact they do not, as an elevation view of Anaximander's cosmos makes clear.[92] The technique he adopted was intercolumnar, not interaxial, though he could have selected that technique. The proof that he did not is that Anaximander's numbers make sense only when we grasp that he employed the intercolumnar technique. The sun wheel is expressed as 27 or 28 *plasiona.* Once we see the model that Anaximander was adopting and adapting, it is clear that the numbers are not "sizes"—they are neither diameters nor circumferences—but rather are *radii.* Thus, the numbers are *distances* to the heavenly wheels, 27 earth-column diameters to the inside of the wheel, and 28 to the outside of the wheel, and each wheel is one module—that is, one earth-diameter—thick. Thus, the number 19 is not a corruption but was in all likelihood identified on the informal sketch that Anaximander believably made and included in his book; the wheel of the moon was 19 earth-diameters to the outside of the wheel and 18 to the inside (fig. 3.5). And the unsupplied numbers of 9 and 10 express the distance from the earth to the star wheel using the architect's modular technique. Now, while it is certainly true that each wheel is surrounded by moist air of an undetermined thickness, it is irrelevant to Anaximander's calculation once we grasp that he was imagining the cosmos using an intercolumnar, architectural technique. This is why the ingenious speculations by O'Brien[93] are mistaken because, while he did not see the architectural connection, he still explains his own hypothesis as if Anaximander's technique entailed what amounts to interaxial measurements.

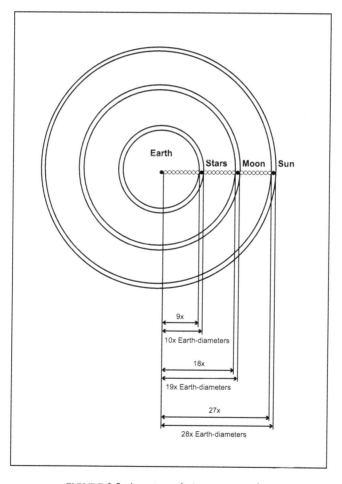

FIGURE 3.5. Anaximander's cosmos in *plan*.

To be sure that we understand how different the cosmos appears in elevation, I present a connected series of elevation views in figure 3.6. Of course the most formidable problem is, again, to ask how Anaximander imagined the cosmos, from an imagined position outside the cosmos. From where was he "as if standing"? It seems that he would have had to imagine while somehow standing *outside* of the cosmos. Had he imagined the cosmos as the architects certainly imagined their temples, while seeing them across and from afar, the following illustrations may prove useful. In any case, the key point is to observe how very different are the images of the cosmos in elevation. Could the reader have easily imagined the elevation view of the cosmos from the plan alone, and the plan from the elevation? The elevation view containing the wheels of sun, moon, and stars with the column-drum–shaped earth in the center is presented at the far right. The outermost wheel of the sun, and the virtual cylinder on which it rises and falls during the course of the year, without changing its angle of inclination, is presented on the far left. The virtual cylinder that contains the wheel of the moon is second from the left; the virtual cylinder that contains the wheel of the stars, third from the left, must stretch as if infinitely far, like looking down railroad tracks that seem to continue to infinity, because there are no portions in the sky where the stars are absent. In the elevation view, we can see already the structure of a cosmic axle, suggesting an *axis mundi*.

By appeal, then, to architectural techniques that deal with column drums and column placement, we have one line of argument from archaeological resources that proposes to resolve the scholarly debates about the cosmic numbers and the earth's placement as *homoiôs pros ta eschata echon*. The archaeologists provide the evidence for archaic architectural techniques that Anaximander observed and imaginatively employed. But the archaeologists have another argument, useful in their own research, that lends additional, separate support to this position on Anaximander's cosmic numbers. To that argument—technological style—we now turn.

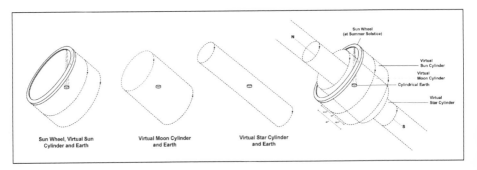

FIGURE 3.6. Elevation views of Anaximander's cosmos. The virtual cylinders that contain the wheel of the sun (far left), moon, and stars, finally incorporated together (far right).

C.2

The Archaeologists' Idea of *Technological Style*
Sarah Taylor, co-author

The appeal to technological style[94] offers the archaeologist a way to understand how the cosmic numbers of 9/10, 18/19, and 27/28 were likely Anaximander's selection since they follow the archaic formula of 9 and 9 + 1; this is a completely different argument from the one we just considered. The appeal to the architect's intercolumnar techniques supplies a way to understand how Anaximander imagined cosmic distances. The appeal to what the archaeologists call technological style, however, helps us understand Anaximander's specific selection of cosmic numbers. When we grasp this archaic formula, we have additional evidence, then, for supporting the view that Anaximander's numbers describe, as Hesiod described cosmic distances as 9 + 1 anvil days, distances of "far, farther, and farthest" on a completely extraordinary scale. Not based on astronomical observations, and perhaps in some sense unconsciously selected, Anaximander appealed to an archaic formula known to all of his contemporaries. While Hesiod expresses cosmic distance to the highest and lowest depths of the cosmos using the formula of 9 and 9 + 1, Anaximander ventured to claim that there was "space," a series of enormous depths in the cosmos. While the stars were immeasurably far, that is, 9 + 1 modular units away from earth, the moon was + 9 modular units farther, and the sun + 9 modular units farther than the moon. Let us consider how archaeologists justify a range of proportional inferences by appealing to the argument for technological style.

Archaeology is a discipline that relies almost exclusively on material culture to reconstruct past social systems and ways of life. There are many ways to view material culture, and each perspective provides different kinds of information that may help archaeologists in achieving their research goals. Style and material culture have a long history in archaeology and continue to be debated to this day. Aside from the aesthetic and immediate functional values of material culture, each object contains information about the maker, the user, the available resources and attitudes toward those resources, as well as information about the technological values and knowledge (and how those ideas interface with other aspects of society), and broader ideological concepts of the culture in question.

Dishes, for example, are primarily (but by no means exclusively) utilitarian or functional objects that we use in our daily lives. Dishes can be made out of a variety of materials, including, paper, ceramics, porcelain, metal, plastic, glass, or wood, all of which serve the function of a dish with equal utility. However, each of these materials carries different implications when fashioned into a dish. We tend to use porcelain dishes on special occasions, glass for daily meals, and plastic or paper at picnics, parties, and other informal gatherings. Wooden dishes are associated with salads and earthenware dishes are usually used for coffee and soup. Metal dishes have a variety of uses ranging from aluminum dishes for camping to silver tea sets that are rarely used and often displayed. The context that different materials are used in is intimately related to the way we perceive these materials in relation to the function of serving and consuming.

The value of a dish is dependent not only on the material of which it is made but also the technique used to make it, for those very attributes influence the way the object is perceived. Because individual craftsmanship is not common in our daily lives today, and because of the time and energy invested in such production, a handmade ceramic platter is more highly valued than one made by a machine. Furthermore, there are size-related ideals as well. The sizes in which a given dish is made usually reflect the normal amounts of the appropriate food/drink that either one person or a group of people consume. The quantity of a food or beverage that an average person should eat is reflected in the average size of single serving dishes. Nowadays, one would be surprised and amused to be served beer in a porcelain teacup.

The point is that the archaeologist proceeds on the reasonable assumption that material culture contains more information than most people realize. It is not enough to use complex chemical analysis to identify the source of the material of which an artifact is made, the way it was made, and the form it takes. One should also try to understand why that source and material were used, why certain construction techniques were employed, and why the object takes the form it does. Archaeologists might ask why beer should not be served in a porcelain teacup, but instead in a glass mug. We ask ourselves why it seems comical for a teenaged boy to own porcelain, or why we might be equally amused if the beer at a bar had been served in teacups.

Our concern here is with some of the approaches used to identify the social relationships and values that underlie what we see in material culture. But we are also concerned with the question of the directionality of influence between material culture and ideology, especially cosmology. Among the available approaches toward material culture are two closely related theoretical perspectives: *technological choice* and *technological style*. Because the two approaches are so similar, a discussion of both is appropriate. Let us turn briefly to consider the intellectual history and practical application of technological style, as it related to the work of the archaeologist.

"Style" has been used in a variety of ways throughout the history of archaeology. The very meaning of the word has changed in the course of decades of debate. In the early years of the culture historians, "style" was used to define cultural and temporal boundaries. We all recognize that style, in popular terms, changes from generation to generation, and varies from one region to the next. This truism is the principle behind temporal sequences and cultural distinctions. Culture historians like V. G. Childe and

A. V. Kidder were concerned with creating typologies and seriation sequences to help place artifacts in relation to each other temporally, and wanted to define culture groups on the basis of similarities or differences in the style of their material culture.[95] Variation in the archaeological record was accounted for by cultural or temporal differences that were defined on the basis of stylistic differences. Similarities in style were explained as diffusions of ideas from one place to the next. Artifacts were viewed as the product of shared cultural ideas.

As new and improved dating techniques developed, the use of seriation for dating became secondary. Many styles that were thought to represent temporal differences turned out to be contemporaneous. Furthermore, the use of style as a cultural identifier alone came to be seen as a static approach to dynamic human behavior. By the 1940s, some archaeologists, including Childe, began to question the idea that ethnicity or cultural groups could be defined archaeologically, and that diffusion could adequately account for many of the similarities in the archaeological record.[96] In response to these criticisms, the discipline turned toward a variety of functional and contextual interpretations of material culture that are jointly referred to as processual archaeology. Among processualists, style was viewed as an adaptational response to environmental or economic constraints. Religious or ideological explanations for style as an adaptive response were especially popular. The purely synchronic tendency of many functional interpretations failed to account for internal change through time, and the environmental determinism of neo-evolutionary approaches disembodied individual ingenuity from the history of technology. Many archaeologists sought a more behavioral interpretation of material culture that would explain cultural processes.

In the 1960s, some archaeologists, such as J. Caldwell, started using style as a marker of ethnic boundaries and social groups through time.[97] Others, like Binford, maintained that while style could reflect ethnic identity, it usually reflected functional difference. Viewing style as a symbolic message that reinforced identity and group values would seem to have taken it out of the static hands of culture historians, only to place it in the passive hands of processualists, where style was used to explain diversity in the archaeological record through functional similarity of the items themselves. Processualists used style to explain cultural processes and systemic change. Truth be told, processualists as a group were never able to agree on the definition or role of style. Iconographic studies of style were countered by isochrestic approaches,[98] that is, invisible and visible approaches to style, and evolutionists like Dunnell[99] constantly debated with behavioral archaeologists, such as Schiffer, over the role of style.[100] The primary focus of most processualists was the system behind the people and artifacts. Artifacts and their styles (traits) were not viewed as the end product of peoples' ideas and perceptions about what that item should look like, but were thought of as representations of one part in a larger system.[101] The idea was to study parts of a system, and later integrate all the parts into a single system. These goals turned out to be too ambitious, and the integration of the parts of a society rarely happened. In other words, most processualists studied only one or two parts of a total social system without integrating even those two spheres. The ideas about style as part of an integrated social system can be attributed to this time period.

In time, many archaeologists were frustrated by the positivist approach of processualists and wanted something different. Furthermore, processualists and culture historians had moved along a similar trajectory toward increasingly divorcing the idea of the individual as a decision-maker from prehistory. In response to the growing dissatisfaction with processualism, a myriad of theoretical perspectives, jointly referred to as postprocessualism, gained popularity. I. Hodder and M. Leone are among the most vociferous postprocessualists and are usually considered the spokesmen for the conglomerate of perspectives. Broadly speaking, postprocessualists view style as an active agent of culture, but also as a social product and social practice. Postprocessualists as a whole were not, however, concerned with the historical processes of culture or style. And while they championed the search for the individual in archaeology, they never achieved that goal. Postprocessualists always placed the individual beneath the overall social structure. In this sense, style itself was more of an agent than the person producing the style.

One theoretical perspective that falls under the broad rubric of postprocessualism is French structuralism. Structuralism merits focus here because it was instrumental in informing the concepts of technological style and technological choice. Structuralism is closely related to French social theory, which was strongly influenced by Durkheim's "superorganic"[102] concept and the work of Mauss.[103] The "superorganic" is a sort of collective unconscious that guides human behavior. French social theorists, like Mauss, viewed symbols as mediators between individuals and society. French structuralism combines French social theory with linguistic theory to search for the unconscious truths that define peoples' perception of the underlying social structure. French structuralists, for whom Levi-Strauss is often considered the spokesperson, believed that they could find universal constants underneath human social structure. By searching for universal laws and unconscious truths, structuralists denied the importance of intentionality and history on the social and cultural process. Structuralists see style as a symbol of the underlying social structure, which is always approached via linguistic analogies with a focus on cultural dichotomies. The underlying structure that structuralists seek is intimately related to the basic cosmological vision of a culture.

Lemonnier[104] criticizes structural anthropology on three grounds: (1) structuralists tend to limit data to the study of artifacts alone; (2) their analyses do not proceed beyond the "de facto" immediate informational content in the artifact; and (3) the informational content is only viewed from the perspective of the role it plays in conflicting social relations.[105] By looking for universals, structuralists fail to recognize the uniqueness of cultures. And because these universals are derived from structural elements of a culture, which usually include a basic worldview or cosmology, the structural elements themselves are simplified to the point of near uselessness. What archaeologists need to do is to try to relate structural elements of technology to other social relations.[106]

Postprocessualism has been somewhat eclipsed by postmodern theories that emphasize the role of history and the individual in cultural processes. Postmodernism started long before postprocessualism lost popularity; in fact, some argue that postprocessualism is a kind of postmodern perspective. In general, postmodern social theory focuses on individual agency in society and self-reflective critiques of anthropology. Alongside postprocessualism, but conceptually more closely related to postmodern theory, the anthropology of technology and related ideas developed, one of which is the concept of technological style.

Technological style is a concept closely associated with the theoretical perspective known as technological choice. Both technological choice and technological style are by-products of French structural anthropology of the 1970s. However, they have various conceptual parallels with the sociology of technology and the work of Cyril Smith in the 1960s.[107] The most immediate difference between the two is that technological choice is a French tradition (the *techniques et culture* school of thought and the anthropology of technology), while technological style is an American one; however, there are more important differences that will be discussed shortly.[108]

Heather Lechtman coined the term "technological style" and can be regarded as the primary spokesperson for the concept. Her conceptualization of technological style was influenced by the "humanistic" vision of technology advocated by C. Smith, and grew out of French structural anthropology. Smith held that systems, whether they are biological or social, are made of relationships between parts.[109] It is the relationship, as opposed to the parts, that reinforces the larger social pattern, and it is our understanding of the formal arrangement of relationships that we consider style. Stated differently, the social representations that underlie technological style represent the perspective of the producer toward the materials, and the way the community at large views the nature of the technological action and the product of that action.[110] As a member of the community, the way the producer views the technological event and product, and the way the community at large perceives the materials, are also part of the matrix of social representations that underlie technological style. For example, during the Early Bronze and Early Cycladic Periods in Greece, ceramic vessels found in graves were often poorly fired while vessels in other contexts are all well fired.[111] This may demonstrate a desire to conserve fuel and energy by firing burial goods at a lower temperature. Because burial goods do not remain in circulation and were frequently produced for the burial ceremony, fuel may have been considered too valuable to use in large quantities on burial goods. While low-temperature firing may suggest a lesser time investment in the production of burial goods, the forms of the vessels do not. Burial vessels during these early periods were unmanageably large and heavy with elaborate forms, fine slips, and beautiful polishes. This data could reflect a desire to minimize fuel consumption by firing at low temperatures, but to deliver unusual, perhaps ritual, vessels to the grave. This case study and the previous description of technological style explain Lechtman's early vision of the concept. In this form, there is little or no difference between technological style and technological choice, viewed from a materialist perspective. So, what then is "technological choice"?

Technological choice as defined by Pierre Lemonnier is the "process of selection of technological features invented locally or borrowed from outside."[112] The main difference between this approach and those preceding it, is that technological choice is more about the process of selection than about what is selected. Technological style was influenced by social technology that, like current social theory, is about the role of material culture in the production and reproduction of social relations and values,[113] but it was influenced by French social theory to a greater degree. The question then becomes: What influences the choice of technique and where do we look for these choices? Lemonnier asserts that we must look for the social representations of human action on the material world that are embedded in other representations.[114] One must consider whether these choices are real or the work of the anthropologist. One must also ask, as Lemonnier does, whether these choices affect the style or the function of an artifact, what aspects of technology are part of the system of meaning via their physical characteristics, and what the role is of arbitrary choice. Thus, the technological-choice approach examines the process of technique selection on both the population and individual level.

While the technological-choice approach does not disagree that techniques and technology have an effect on matter and that material culture is often symbolically charged, these aspects of technological studies are downplayed in an effort to counterbalance the persistent focus of earlier material culture studies on the symbolic aspect of shape and decoration as communicators of social messages. This is really how technological choice and technological style differ. Technological choice looks principally at technique on its own, rather than seeing only the relationship between technique and society, but also regards how it relates to the relationship between technology and culture,[115] while technological style includes the symbolic aspects of material culture. Of particular importance is the concept of embedded technology that concerns how choices are made, reproduced, and maintained, as well as how they fit into the larger structure of a given society and how access to resources is both justified and maintained.[116] In this respect, technological choice combines the goals of earlier schools of thought, while refocusing the analysis. Lemonnier was always interested in the social side of technology; Lechtman came to be interested in this side of technology only as her work developed. The two approaches have a somewhat parallel development, though the French tradition is much older. What about technological style today?

In time, Lechtman grew more interested in examining the underlying ideas beneath material culture. It is at this point that technological style most clearly distinguishes itself from technological choice. When archaeologists use the term "technological style," they are referring to Lechtman's later concept of the term. The objection might well be raised that studies of archaeological materials that claim to be using the concept of technological style remain mired at the point where technological style is still the same as technological choice—the difference between the two being the degree to which the social is emphasized. In her later work, however, Lechtman compared technology to the visual arts, emphasizing the way that both can reflect cultural ideas or values that are part of the action itself.[117] For example, a ballet recital may reflect numerous cultural values, not the least of which is the viewing of the recital itself and the choice of movement and aesthetic body form. It would also reflect values related to the costuming and set design (not to mention the materials chosen for those elements), or the cultural ideas expressed in the stories and music specifically selected. Likewise, the process of making bread reflects many cultural values, including the cultural ideals associated with the making and eating of bread, the use of specific ingredients and tools, the kind of oven used to bake the bread, what persons in a culture perform this action (and which do not), when it is appropriate to make and consume bread of what types, and so on. The difference between technological style and technological choice can furthermore be seen in the material lens through which they view social perceptions. Technological style is principally archaeological while technological choice is primarily an ethnographic technique.

Thus, *technological style* can be defined as the "formal extrinsic manifestation of intrinsic pattern."[118] For example, despite the fact that the potter's wheel had been in use for several centuries, classical Greek burials often contained handmade ceramic vessels alongside wheel-thrown vessels, though handmade vessels have not been recovered outside of a burial context.[119] This example suggests that handmade pottery held some kind of intrinsic value that was associated with burials. It may be that handmade crafts were highly regarded for their craftsmanship and were therefore desirable burial offerings. Whatever the case, the two formation techniques were viewed differently by the community at large, even though the end product may have differed very little.

As Lechtman defines it, technological style is the cultural patterning that cannot be cognitively known by members of a culture, but that influences the choice of technological technique.[120] While technological style is usually discussed in terms of unconscious influences or patterns, not everyone believes that those cultural patterns are necessarily unconscious influences.[121] Unconscious patterns can be understood most easily in terms of tradition or habitus. Indeed, technological style, as both Stark[122] and Dobres[123] have noted, is the technological expression of habitus. For example, if I were shopping for a girl's dress to give a friend at a baby shower, I would most likely buy a pink dress. While I would know why I was choosing a pink dress (because pink is a color associated with little girls and therefore most appropriate), I could not tell you *why* pink is associated with little girls. Furthermore, I would be most likely to buy a pink dress because most of the baby dresses available for purchase are pink, a reflection of the cultural pervasiveness of this association. Habitus, as defined by Pierre Bourdieu, is "the universalizing mediation which causes an individual agent's practices, without either explicit reason or signifying intent, to be nonetheless 'sensible' and 'reasonable'"[124] There are any number of ways of doing something and any number of possible or available times to do them, but one way and one time will be considered better than the others, though there may be no discernible logical reason that it is. This definition implies that logic is not a part of the habitus when, in fact, even logic is culturally and historically determined.[125] Habitus is history turned into nature—the unconscious quality of habitus is history forgotten.[126] An example of unconscious patterns behind techniques viewed as habitus might be the appropriate order of preparing a meal. In modern Greece, for example, one must always thaw a roast before putting it into a pot and cooking it or you will elicit disapproval of your cooking in the neighborhood.[127] The use of time-saving kitchen appliances in the cooking process, as Sutton has showed, is viewed with suspicion and distaste. Food prepared with these devices is called "prostitute food" as it suggests that the woman was saving time on the preparation process in order to do something else. Sometimes, however, it seems we are conscious of the cultural patterning behind technological style. For instance, when one does something in a given order or makes something with particular materials for personally ritualistic or sentimental reasons, or does something as a conscious act of defiance, that person has acted within his or her social and historically constituted belief system, consciously. For example, one might consciously build a barn and feed the livestock the same way, in the same order that one's grandfather did, not because it is traditional or better than other ways, but because it reminds one of him or because the person wants to honor his memory. Craftsmen, in particular, are frequently conscious of the reasoning behind the choices they make.[128] Lemonnier's use of the term "representation" often refers to those same kinds of unconscious choices.[129] For Lemonnier, these unconscious choices, or technological styles, can lie at multiple levels of association. For instance, a given material may be associated with gender/age/class, and so on, with a particular technique or technological action, and there may be a logical relationship between aspects of a given material and aspects of other materials. Lemonnier questions why a given aspect of technology is used to express certain social relations while another is not.[130] For example, why does the design on a vessel convey ethnic identity, as opposed to the shape or material composition of the vessel? Also, technologies and technological style involve an element of arbitrary decision and explicit choice that are not related to the immediate physical constraints of the material or design, but do represent "higher" levels of systems of meaning in a given society. Style, then, can be seen as a sort of "technological performance" of mental schemata that do not necessarily contain any immediately observable meaning, but can be learned and passed on from each generation to the next.[131]

Now that we have surveyed the archaeologist's exegesis, let us apply this discussion to Anaximander's selection of a column drum to represent the size and shape of the earth, and to the problem of the cosmic numbers. Let us first turn to review the ideal of a column as an example of technological style. Most especially, let us review the proportions of Ionic columns in archaic Greece. While it is fair to say that Gruben pointed out some of the problems of precisely identifying Ionic column height and proportions,[132] there are certain recurring features. Ionic columns were more slender than the Doric ones erected contemporaneously. While Doric proportions are often four, five, or six times the height of lower column diameter, the Ionic ones were thinner, and, based on surviving examples from Delphi, have been generally reckoned to reach a height of nine or ten times the lower column diameter.[133]

The column had symbolic meaning for the archaic Greeks.[134] A case has already been proposed by Schaus, Yalouris,[135] and others that identifies Anaximander as a proponent of this view. In addition, much supporting evidence, as we have already considered, comes from the appearance of columns and column drums on painted vases. The symbolic meaning of the column has also been argued for at length in other studies,[136] and, based on Anaximander's claim that the earth resembled a column drum, we have far-reaching support for this particular symbolism.

This recurrence of the formulas of 9 and 9 + 1 shows that it was widely understood throughout the archaic Greek world, and lends credence to the hypothesis that Ionic temples had colonnades whose column heights were likely 9 or 9 + 1 times the lower column diameter, as a matter of technological style. This same proportional value also appears in poetry to connote greatness in size, scale, or distance. In the opening of Homer's *Iliad*, we learn that the war is in the ninth year and will end in the tenth; in the *Odyssey*, Odysseus labors for nine more years to reach home and arrives in the twentieth. Here we see the recurrent formula of 9 and 9 + 1 to connote great amounts of time, and the formula can be additionally compounded for effect. In the *Homeric Hymn to Demeter*, we are told that Persephone wanders the earth for nine days looking for her mother. In Hesiod, when the poet wants to suggest greatness of numbers, he also appeals to such aggregations as nine swirling streams. When Hephaistos was thrown from heaven, he fell into the sea and was rescued by Thetis and her sister Nereids; they kept him with them in a cave for nine years, indeed a long time.[137] In astronomical size, Hesiod explains that the height of the heaven is so great that, if one were to drop an anvil from the height of heaven, it would fall for nine days and nights before reaching the earth on the tenth; the symmetrical distance of the depths below the earth is also measured by 9 + 1 anvil days from Hades to Tartarus. Clearly, 9 and 9 + 1 provide a pervasive formula for expressing great size, scale, and distances in Ionian culture. The formulas 9 and 9 + 1 are examples of what the archaeologists call technological style. It might be that, like the selection of a pink dress for a baby shower for the soon-to-be-born daughter, neither Homer nor Hesiod could easily explain, if asked, why they employ the formulas of 9 and 9 + 1, but what they knew most assuredly was that their audience, their contemporaries, would have grasped immediately that great amounts of time, numbers, and distances were being suggested.

In terms of temple building proportions, both Doric and Ionic architects built temples with similar materials, in similar ways, and for similar functions. The proportions of the temple elements, however, differed. This was clearly not because their aesthetic preference for ratios differed; the 9 or 9 + 1 ratio was chosen deliberately by Ionic architects, according to the argument before us, for reasons of *technological style*. Since the process of production is an integral part of technological style and must be part of any example of it, the question of whether and to what degree these formulas appeared in sculpture, music, or other arts would have to be addressed in a full-length study of Ionian technological style.[138]

In the context of technological style in archaic Greece, then, debates about Anaximander's sizes of or distances to the heavenly wheels are presented in a new light. Once we see the archaic formula, then the cosmic numbers appear as iterations of the 9 + 1 and 9 formula. The appearance of this formula in Anaximander's map of the cosmos draws a clear analogy between the earth and the column drum. A key part of Ionian architecture, something for which Ionians are remembered, as Finley rightly observed, was the conscious planning of what a temple would look like when finished.[139] Ionians imagined the building before building it. Anaximander also imagined the cosmos before building a map or model of it. Therefore, imagining the final product before building it and incorporating the pervasive 9 and 9 + 1 ratio/value appear to be part of an Ionian technological style.

Up to this point, we have provided a basic description of how some archaeologists infer abstract concepts from material culture. We have also mentioned the theory of structuralism and how it seeks to identify a world-view or cosmology via linguistic analogy, as well as how structuralism informed the development of technological style. In particular, we have explored the historical development and ideas behind technological style. It is now clear that, from the archaeologists, we learn that both products and processes of production are influenced by pervasive cultural ideas and perceptions. These ideas and perceptions enter into material culture in ways that may be largely, though not totally, unconscious to the producers themselves. It is also clear that perceptions and ideologies, especially cosmological ones, are cultural phenomena, and thus archaeologists do not regard them as inherent truths. If this is the case, then we can agree with structuralists in claiming that myth, ritual, and cosmology reflect wider social ideals just as does material culture. Few archaeologists would disagree with these generalities. What is missing here, however, is an approach toward understanding the relationship between material culture and cosmology, with the craftsmen as intermediate to both. Can both material culture and cosmology be seen as the products of a kind of craft production? In other words, is thought a craft?

Archaeologists who would deny that "abstract, speculative thought" is a craft might argue for this position because the construction of the universe is not economically or functionally valuable. Simply stated, it is not a tangible product. Priests, philosophers, theoreticians, and politicians are all recognized as specialists in the world of archaic Greece, just not as craft specialists. And yet the construction of a cosmology reflects already existing values and analogical relationships between things and ideas. For example, Anaximander chose to draw on architecture in conceptualizing the cosmos. The thinker could have chosen any analogy. Why architecture? If technological style is defined as the processes and products of craft production that embody more pervasive social perceptions and structures, then the universe as a cultural product is just as much a subject of technological style as the dish or pot. On the other hand, if technology is defined as *our* means of interacting with the environment, which a dish or pot certainly do, it seems that cosmology arguably does not.

This issue must be left open-ended here for the reader to contemplate, and it is a suitable subject for future study. It most certainly would raise a variety of responses from the archaeological community. Some might agree hesitantly that such an application is possible; many more might emphatically deny one. The question is whether technological style can be expanded to include "cosmology"—our production of a universe. This question does not appear to be routinely asked or answered by archaeologists.

Nor does the important issue of "directionality" seem to be addressed in discussions about technological style. That is, do already existing values and perceptions influence the production of material culture, or do the processes and characteristics of craft production influence the broader social values and perceptions? It would seem that both are true to some degree; the relationship is certainly an ongoing dialectic between the social world and technology. However, determining the directionality of a given case study is beyond the scope of archaeology at this time. Generally, technological style is about the way *culture influences technology*. The study here on Anaximander suggests the importance of a dimension whereby the way *technology influences culture* deserves a fuller hearing.

In summary, what archaeologists will likely agree to is that material culture is directly linked to underlying social values and that the perceptions associated with matter influence the form-material-technique of a given object. "The relationships among the formal elements of the technology establish its style, which in turn becomes the basis of a message on a larger scale."[140] It seems that there is a good deal of ambiguity and personal modification/interpretation of technological style as the concept is applied by archaeologists. Careful reading of Lechtman's use of the term provides the following archaeological guidelines. "Technological style" combines the function, style, and production process of an object. These combined elements are a technological style. Technological style can be explored through looking at the choices about materials, techniques, form, and so on made during the production process. Lechtman believes these choices are unconscious, while others disagree. The appeal to technological style is a way to illuminate underlying social perceptions that influence multiple aspects of a culture. The underlying concepts identified should be present in multiple mediums or artifact categories to qualify as technological style. Unfortunately, few archaeological studies go this far, and none of the examples considered here was expanded to include other domains. The conceptual production process of the different crafts must be parallel. Studies of technological style must take place within the same culture, during one time period, or sequential time periods. Technological style is not yet used at a regional level.

FOUR

Anaximander's Cosmic Picture

The "Bellows" and Cosmic Breathing

A

THE DOXOGRAPHICAL EVIDENCE

ἁρματείῳ τροχῷ παραπλήσιον, τὴν ἀψῖδα ἔχοντα κοίλην, πλήρη πυρός, κατά τι μέρος ἐκφαίνουσαν διὰ στομίου τὸ πῦρ ὥσπερ διὰ πρηστῆρος αὐλοῦ.

—Aetius II, 20.1 [DK 12A21]

[Anaximander says the sun] is similar to the wheel of a chariot, which has a hollow rim, full of fire, and the fire appears at one point as through a mouth-like opening just like the mouthpiece of a bellows.

Ἀναξίμανδρος [s.c. φησι γίγνεσθαι τὴν ἔκλειψιν ἡλίου] τοῦ στομίου τῆς τοῦ πυρὸς ἐκπνοῆς ἀποκλειομένου.

—Aetius, II. 24.2 [DK 12A21]

Anaximander [says that in the eclipse of the sun] the fire at the breathing-hole is closed up.

Ἀναξίμανδρος [s.c. φησι τὰ ἄστρα εἶναι] πιλήματα ἀέρος τροχοειδῆ, πυρὸς ἔμπλεα, κατά τι μέρος ἀπὸ στομίων ἐκπνέοντα φλόγας.

—Aetius II, 13,7 [DK 12A18]

Anaximander [says of the heavenly bodies that] they have the form of wheels, formed by compressed air and filled with fire, that exhale flames at one point through a mouth-like opening.

τὰ δὲ ἄστρα γίνεσθαι κύκλον πυρὸς ἀποκριθέντα τοῦ κατὰ τὸν
κόσμον πυρός, περιληφθέντα δ᾽ ὑπὸ ἀέρος. ἐκπνοὰς δ᾽ ὑπάρξαι
πόρους τινὰς αὐλώδεις, καθ᾽ οὓς φαίνεται τὰ ἄστρα. διὸ καὶ
ἐπιφρασσομένων τῶν ἐκπνοῶν τὰς ἐκλείψεις γίνεσθαι.
—Hippolytus, *Ref.* I.6.4 [D-K 12A11]

The heavenly bodies come into being as a circle of fire, separated off
from the original outer fire, and each circle is encased in air. There are
some pipe-like passages that form breathing holes at which openings the
heavenly bodies appear; consequently eclipses occur when these breath-
ing holes become occluded.

B

THE SCHOLARLY DEBATES OVER
THE TEXT AND ITS INTERPRETATION

THE SCHOLARLY DEBATE over the text and interpretation of *prêstêros aulos*
(πρηστῆρος αὐλός) is both curious and pivotal. What is at stake in constru-
ing this technical parlance is an understanding of *how* the fire that constitutes
the sun, moon, and stars makes its appearance. The expression *prêstêros aulos*
holds the promise of revealing the mechanism for, or providing the explanation of, how the heavenly fire, concealed by compressed mist, nevertheless is
able to become visible to us. The sun, moon, and stars are not heavenly bodies as such, but rather are likened to wheels of a vehicle of transport—perhaps a chariot, cartwheel, or the archaic architects' invention of wheels that
transported the blocks from the quarry to the temple building sites—their
rims are hollow and full of fire. The fire appears through an opening in the
wheel—*hôsper dia prêstêros aulos*.[1]

A review of the scholarly debate reveals two approaches, both of
which supply distinct *mechanical* explanations; I shall call the first an
organic approach and the second a *meteorological* approach, though the
matter has never been classified this way. The problem emerges because
the expression *prêstêros aulos* is a ἅπαξ λεγόμενον, a "once occurring
phrase" in which the technical parlance of *prêstêr* (πρηστήρ) and *aulos*
(αὐλός) appear together for the first time in the surviving corpus. There is
a conventional reading that suggests what I shall call the *organic* approach,
traceable to Diels who first suggested rendering *hôsper dia prêstêros aulos* "as
through the nozzle (or mouthpiece) of a bellows." There is another
approach, hinted at by other scholars but most recently argued for at
length by Couprie, that I shall call the *meteorological* approach, in which
hôsper dia prêstêros aulos is rendered "as a permanent jet or stream of lightning." Both interpretations supply a *mechanical* explanation of *how* the fire
in the wheels makes its appearance to us.

The conventional translation of the phrase can be traced to the great scholar Hermann Diels, who explains in his *Doxographi Graeci*: "*immo prester est follis fabrorum*";[2] in his German text, *Die Fragmente der Vorsokratiker*, he translates '*Blasebalgröhre.*'[3] Thus, Diels' translation is "as through the nozzle (or mouth) of a pair of bellows," an expression he attributes directly to Anaximander himself. The list of scholars who accept or adopt this translation is extensive, and includes Burnet,[4] Freeman,[5] Kirk-Raven,[6] Guthrie,[7] Fränkel,[8] Brumbaugh,[9] Robinson,[10] de Vogel,[11] West,[12] and others. Several of these scholars, along with Kahn[13] and Lloyd, acknowledge and discuss the difficulty of this translation of *prêstêros aulos*. Lloyd concludes a long note in which he acknowledges that while the interpretation in Aetius is obscure, "'bellows' is at least etymologically possible as a meaning for πρηστήρ, and this appears to give the best sense for this difficult phrase."[14]

In the secondary literature, however, what I am calling the *organic* case that is suggested by the image of the bellows at work has been alluded to but not elaborated upon. Stated succinctly, the *organic* case is this: Anaximander is a hylozoist—the whole cosmos is alive; it is alive by "breathing," through a variety of stages of inhalations and exhalations. There is an ultimate under-lying stuff from which all things derive that Anaximander called *to apeiron*[15]—and Thales called *hydôr*[16] and Anaximenes called aêr,[17] though this conventional interpretation of Milesian "material monism" has recently been challenged by Graham.[18] This originating stuff appears as earth, water, air, and fire, and intermediary stages of these material forms. Water evaporates, for instance, due to the heat of the heavenly fire, that is, water *exhales* as mist; "evaporation" is "exhalation." The evaporating moist air rushes upwards and surrounds the heavenly, fiery wheels, where it compresses and thus conceals them. The moist *aêr* "nurtures" or "feeds" the heavenly fire as an *inhalation*, and the fire in turn radiates from those heavenly wheels as an *exhalation*. Thus, the heavenly part of the cosmos does not breathe air, as we do; in the heavenly exchange and transformation of materials, the cosmic wheels inhale nourishing moist air and exhale fire, that is, breathe fire. The translation "as through the nozzle of a pair of bellows" conjures the breathing mechanism of inhalation and exhalation. The radiating light from the fiery wheels, then, is cosmic exhalation. Freeman's translation precisely captures this sense when she renders Aetius II, 13, 7 *Anaximandros* [sc. *phêsi ta astra einai*] *pilêmata aeros troxoeidê, puros emplea, kata ti meros apo stomiôn ekpneonta phlogas:*[19] Anaximander [says that the stars are] "pads of air, wheel-shaped, full of fire, *breathing out flames at certain points through mouths.*"[20] Guthrie also translates "breathing out flames," and for *ekpnoas* in Hippolytus I.6.4, he translates "breathing-places."[21]

In opposition to this reading of *prêstêros aulos*, Couprie has recently argued that the conventional view cannot be adequately defended, and he summons comments from other scholars who have given voice to what I am calling the *meteorological* explanation in the process of building his thought-ful argument. Couprie urges us to replace Diels' translation with something like "a permanent jet, beam, or stream of lightning." Thus, Couprie argues that *prêstêros aulos* has a meaning directly connected to meteorological phe-nomena, and the mechanism by which fire radiates from the heavenly wheels is the same by which Anaximander seems to have explained how lightning appears in a storm, as a result of compressed air. The strength of this case rests largely on appeal to Aetius,[22] who when referring to thunder, lightning, thun-derbolts, whirlwinds, and typhoons, claims that Anaximander attributes all these meteorological phenomena to winds; when wind becomes compressed in thick clouds, it bursts out making a loud noise, and in the forced rupture a flash of light is seen against the black clouds. The *meteorological* case, then, is that the appearance of heavenly fire is attributable to the same mechanism as lightning, and that is the result of compressed air. Thus, Couprie takes *prêstêr* to mean "lightning fire" and *aulos* to be construed as a "jet or continuous and permanent stream." How shall we decide the matter and thus answer the question *how* the heavenly fire appears to us?

The word *prêstêr* often occurs in passages that relate powerful meteoro-logical phenomena:[23] powerful winds accompanied by heavy rain and light-ning. In Hesiod's *Theogony*, the expression *prêstêr anemôn*[24] appears; West notes that it is an unusual reinforcement of Zeus' thunder and lightning, emphasizing the burning element.[25] Perhaps the translation "scorching winds" adequately conveys that sense.[26] The word *prêstêr* also occurs in Euripides and shows it can have the meaning of a "jet" or "stream" of blood.[27] The word *aulos* also has a range of meanings. It most commonly refers to wind instruments such as a pipe or flute, even a clarinet or trumpet. However, *aulos* can also mean simply a "hollow tube, pipe, or groove"; in the *Iliad*, it refers to a "tube" or "socket" of a spear into which the point of a lance is fitted;[28] in the *Odyssey*, it is used to refer to a jet or stream of blood *through the tube of the nostril.*[29] In the Hippocratic corpus, the expression *aulos ek chalkeiou*[30] may be rendered explicitly "the blacksmiths' bellows."[31] And the word *aulos* also appears in Aristotle's *Problemata*, meaning the "tube" of the clepsydra; in his *Historia Animalium* and the *de Partibus Animalium*, *aulos* refers to the "blow-hole" of an animal, and Aristotle calls upon *aulos* to refer also to the funnel-shaped form of the cuttlefish.[32]

Thus, among the range of meanings that *prêstêr* has assumed, we should consider powerful winds often attended by lightning, a jet or stream like that of lightning but sometimes applied outside meteorological phenomena such as is the case with blood: *prêstêr*, then, suggests a powerful emission that moves in the form of a direct jet or stream, and it can imply scorching air. And among the meanings of *aulos*, we must consider a tube-like vessel that conveys blowing air, identified at times with a blacksmith's bellows, but capa-ble of referring to the blowhole of animals (and even the funnel shape) and hence suggesting a respiratory function. Now, what happens when Anaximander places together these two terms for the first (and only?) time in the surviving corpus: *prêstêros aulos*?

Aetius reports that, according to Anaximander, we see the light of the sun *hôsper dia prêstêros aulou*, that of the moon *hoion prêstêros aulon*,[33] and the light from the stars *kata ti meros apo stomiôn ekpneonta phlogas*.[34] Diels defends his translation by appeal to two different arguments; in the *Doxagraphi Graeci*, he appeals to a passage from Apollonius of Rhodes; in the *Vorsokratiker*, he appeals to two passages in the Hippocratic corpus. In both cases, he reaches the conclusion that the right translation of *prêstêr* is *Blasebalgröhre*, In the *Wortindex*, edited by Kranz, the translation *Blasebalg* is listed.[35] The cautions and objections to Diels' interpretation, however, seem to be threefold: (1) there are conceptual objections, (2) there are objections about the philology, and (3) there are arguments that suggest that the appearance of cosmic fire should be seen as a meteorological phenomenon and thus closely akin to lightning. Couprie argues against translating *prêstêr* as "bellows"; he claims that the image of the bellows obfuscates rather than illuminates the meaning of the text. He follows Lloyd, who pointed out that the bellows blows air into the fire and not fire into the air.[36] Burnet,[37] and Lloyd after him, had wondered also if a meteorological sense might be a more appropriate rendering for *prêstêr*, but Burnet and Lloyd finally leaned away from the meteorological approach and toward Diels' assessment. Burnet concluded that *prêstêros aulos* is simply the mouthpiece of the blacksmith's bellows and "has nothing to do with meteorological phenomena."[38] Lloyd sided with Diels because the meteorological construal of *prêstêr* becomes untenable when placed together with *aulos*. Lloyd argues that "if πρηστῆρ is taken in a meteorological sense in Aetius II.20.1 and 25.1, it seems impossible to interpret the term αὐλός which is twice used in connection with it, for this must surely refer to some sort of pipe or tube and this seems quite inappropriate to such phenomena as lightning or whirlwinds."[39] The central issue at stake, then, is whether the mechanism of meteorological phenomena is the same mechanism by which the heavenly fire radiates. The best case for the meteorological approach would be to argue that compressed air supplies the mechanism that makes lightning and thunder appear, and thus also, mutatis mutandis, the heavenly fire. The greatest obstacle for the meteorological approach is to overcome the objection that while lightning and thunder result from the winds—that is, moving air—being compressed by moist air, the sun, moon, and stars are themselves wheels of fire, not air. The most formidable objection is that these phenomena are not produced by the same mechanism at all; the mechanism by which moist air "nurtures" heavenly fire, and in turn radiates as cosmic exhalations, is not the same by which the winds are closed off by moist, dark clouds that exhale lightning, thunder, whirlwinds, and typhoons. Thus, when Anaximander placed together *prêstêr* and *aulos*, he had some mechanism other than a meteorological one in mind. If Lloyd's objection holds, then the meteorological approach seems untenable when *prêstêr* is placed together with *aulos*.

Lloyd pointed out rightly that the usual word for the "bellows" from Homer onwards is *physa* (φῦσα).[40] And he notes also that the only parallel that can be cited for *prêstêr* is the passage in Apollonius of Rhodes: *deutera d'eis Hephaiston ebêsato, pause de tonge rhimpha tupidôn, eschonto d' autmês aithaleoi prêstêres:*[41] "Next she went to Hephaistos and quickly made him stay his iron hammers: the smoke-grimed bellows [*prêstêres*] withheld their breath."[42] Couprie follows Fränkel; he doubts that *prêstêr* means "bellows" in Apollonius' text, but he argues further that, even if this had been the appropriate meaning, Apollonius' purposes were different from Anaximander's. For in Apollonius' context of Hephaistos' forge, every Greek at the time would have known that he meant bellows. Thus, the passage in question gives Apollonius "the opportunity to exaggerate in order to stress that his story is not about an ordinary forge with normal bellows but about the workshop of a god with its huge and impressive bellows emitting a thunderstorm's blast."[43] And Couprie concludes that "in the context of Anaximander's description of the universe, it is not immediately evident that a bellows would play any role in celestial mechanics. *If* Anaximander had meant to compare the light of the heavenly bodies with the nozzles of bellows, then he would have used the ordinary word *physa* and not the word *prêstêr* which every Greek at the time would have understood to denote a violent weather phenomenon in this context."[44]

Couprie's argument has it exactly backwards. Anaximander is describing the fire that radiates from the heavenly wheels. If Apollonius is "exaggerating" to highlight the more formidable forge of the god over and against the blacksmiths' to show the enormity of its power, Anaximander's metaphor is more powerful still, the cosmic fire itself. Had he used *physa*, only an ordinary forge would be conjured. In Apollonius' passage, the writer calls on an unusual and, according to Diels, a then-obsolete usage of an archaic expression to refer to a pair of bellows; in Anaximander's case, the unusual combination of *prêstêr* with *aulos* denotes a bellows blowing fiery air but also, as Diels argues, connotes a mechanism of cosmic breathing.[45] The unusual combination of *prêstêr* with *aulos* would have signified to the archaic community something other than a usual bellows now that it is applied to the heavenly, fiery wheels, and yet supplied a technical analogy from the blacksmiths' workshop to make the mechanism comprehensible. By the same kind of reasoning, Anaximander clarified the size and shape of the cylindrical earth by the technical analogy with the architect's 3 × 1 column drum at the temple building site. The reason Couprie dismisses this reading, and cannot see how the "bellows would play any role in celestial mechanics," is because he never paid attention to—neither acknowledges nor mentions—the doxographical reports that refer to each opening in the heavenly wheels as a *stomion* or *ekpnoê*. The openings are mouths or breathing holes. The cosmos is alive; part of the story by which the cold and wet and hot and dry interact in endless cycles finds expression in the heavenly wheels that draw moist air to them in the process of evaporation and in exchange breathe out fire as cosmic exhalations.

Diels proposed that *prêstêr* comes from the root *pra* or *par* from which *prêsai*,[46] *euprêstos*,[47] and *prêmainein*[48] are developed, and for which there is a proper notion of breathing (*spirandi notio*).[49] Lloyd made the same philological point when he affirmed that *prêstêr* was derived from the verbal root *prêth* and seems to share "both the senses that the root develops, 'burn' and 'blow'"[50] without, however, mentioning a sense of breathing. To see the strength of Diels' insight, it is useful to remind ourselves of a key ingredient in the thought of the early Milesian philosophers: *hylozoism*. Then, perhaps we can see more clearly how "blowing" and "burning" attain a meaning in the sense of "breathing" and thus favor the *organic* interpretation of *prêstêros aulos*.

In 1987, Furley emphasized the key point that "the essence of the Milesian theories . . . can be given in a single word: hylozoism, the doctrine that matter as such has the property of life and growth."[51] Furley's view echoes generations of scholars throughout the last century. Before him, Cornford, Farrington, Guthrie, Robinson, and others had emphasized the same quintessential point. Cornford wrote in 1912 that the Milesian school is identified with the designation "Hylozoistic—the doctrine that 'the All is alive.' The universe 'has a soul in it,' in the same sense . . . that there is a 'soul' in the animal body. We must not forget that the meaning of *phusis*, at this stage is nearer to 'life' than to 'matter': it is quite as much 'moving' as 'material'—self-moving because it is alive."[52] In 1944, Farrington described it this way: "They [later Greeks] called the Old Ionians *hylozoists*, or Those-who-think-matter-is-alive. That means they did not think that life, or soul, came into the world from outside, but what is called life, or soul, or the cause of motion of things, was inherent in matter."[53] Guthrie devotes a section of his 1962 work to the theme of hylozoism and argues against Gomperz,[54] who dismisses the appropriateness of this parlance. According to Guthrie, for the Milesians, "there is no such thing as dead, inert matter," and so while Aristotle criticized his predecessors for not distinguishing matter from a separate cause that moved it, Guthrie observes how Aristotle's objection was off the mark precisely because the early Greek thinkers could not imagine matter without its being alive and in motion. His argument is that the Milesians, and others, held that meteorological phenomena were living things: Aetius recorded the view, after citing the biological arguments by Aristotle, that in the cycle of living interactions, the sperm of animals is moist and the plants are nourished by moisture, adding "that the fire of the sun and stars itself, and the whole cosmos, are nourished by exhalations from water." While moisture was the nutritive element, as fire is the motive element, Guthrie emphasized that "fire is 'fed' by it, in the form of vapor; and so Theophrastus refers Aristotle's words 'warmth itself is generated by moisture and lives by it' to the whole process of evaporation by which the cosmic fire is produced and replenished."[55] Herodotus avers that "the sun draws water to itself";[56] as the passage shows, he means that the seasons and climactic variations emerge from apparent opposites that nonetheless work in concert, one leading to the other, a view that informed early Greek thinkers and is recounted in the doxography of Anaximander's unending cycle of convergence and divergence of cold and wet, hot and dry.

In the only fragment that survives more or less intact, preserved by Simplicius on the authority of Theophrastus, Anaximander posits an originative source, *to apeiron*,[57] from which all things come and into which, eventually, all things return.[58] In between *genesis*[59] and *phthora*,[60] all the different material forms emerge, unite, and conflict, in a myriad of productions and transformations—earth, water, air, fire, and all variations in between—through the oppositions of cold and wet, hot and dry. This is how the cosmos is alive, and the transformations can appropriately be called inhalations and exhalations. It is these material transformations, these oppositions,[61] that "pay penalty and recompense to each other for their injustice,"[62] but ultimately the return is back into the *apeiron*. One gives rise to the other, one season to the other, in mutual concert and antagonism. Guthrie elegantly stated this important point about transformations of one elemental form into another:

> There is a sense in which water (cold and wet) can and does give birth to its opposite, fire (hot and dry). No other meaning can be attached to Anaximander's sentence than that the 'injustice' which they commit consists in an encroachment, say, of fire that by swallowing up some of its rival water, and *vice versa*. It was in fact a common Greek belief, which emerges still more clearly in Anaximenes, that the fiery heat at the circumference of the universe (that is, in the present world-order the sun) not only vaporized the moisture of earth and sea, thus turning it into mist or air, but finally ignited it and transformed it into fire, the process was actually spoken of as the 'nourishment' of the sun by water or moisture.[63]

Anaximander's sun, then, is a fiery wheel whose heat causes evaporation, resulting in the production of a bountiful supply of moist air that is drawn upwards to it and in turn conceals the fiery wheel; these moist exhalations are inhaled by the sun itself and so replenish and fuel it, exhaled as radiant light and heat, as part of the great cosmic cycle. In the minds of the Milesian *phusiologoi* and other early Greeks, "the whole universe is a living organism."[64] How is it that the cosmos is alive? About Anaximenes, Anaximander's younger contemporary, Burnet points out that in the solitary surviving fragment of Anaximenes' book, he shows how much he was influenced by the analogy of microcosm and macrocosm: on the authority of Aetius,[65] we are told of Anaximenes' belief that *hoion ê phychê ê hêmetera aêr ousa sugkratei hêmas, kai holan ton kosmon pneuma kai aêr periechei.*[66] "As our soul which is air, holds us together, so do breath and air encompass the whole world." Thus, Burnet concludes: "The world is thought of as breathing or inhaling air from the boundless mass outside it."[67] The principle of reasoning is one that analogously connects the life of man with the life of the cosmos; both are alive in a similar fashion. As it was expressed later by Kirk-Raven, "The soul, which is breath, holds together and controls man; therefore what holds together and controls the world must also be breath or air, because the world is like a large-scale man or animal."[68] Anaximander is also credited, by Theodoretus, a late source, with investigating the *psychê*;[69] the view that it "is an airy-nature" is fully credible within the context of Ionian thought, and Aetius says so explicitly.[70] Considering the importance that *pneuma*[71] plays in Anaximander's meteorology, it is difficult to deny that Anaximander also held a belief that there was a vital spirit—*psychê*—in men and animals, despite the paucity of evidence. Kahn urges that "we may perhaps go a step further and conjecture that this vivifying *pneuma* was conceived not merely as 'air,' but also as partaking in the active power of heat,[72] a point echoed and elaborated upon by both Diogenes[73] and Heraclitus."[74] The connection between heat and motion—the very power to be alive—is a theme permeating all Greek natural philosophy.

Anaximander writes from the same Ionian worldview as Anaximenes. Macrocosmically speaking, Anaximander's cosmos is also alive, and alive by stages in which one material form is transformed into another by inhalation and exhalation—breathing—as microcosmically we, too, are alive by breathing.[75] But in the cosmic processes of transformation, according to Anaximander's cosmology, the heavenly wheels made of fire are replenished by the moist vapors that conceal them. In turn, they breathe out a cosmic exhalation like scorching wind. To clarify the mechanism of this breathing, Anaximander appealed to the blacksmith's bellows, thus drawing our attention to a technology by which the fire radiates analogously, and by invoking the unusual phrase *hôsper dia prêstêros aulou*, at once conjures the breathing of cosmic exhalations in nature's great cycle of elemental transformations.

Can an appeal to archaeological artifacts and reports clarify or confirm this interpretation? We now turn to see what the archaeologist can tell us about the bellows, the nozzle, and what archaic Greeks like Anaximander and his audience might have seen. And when we have considered this evidence, we shall see that an archaic bellows offered a technical analogy to illuminate the role that the heavenly wheels played in the great cycles of nature. The mechanism of respiration—of heavenly fire-breathing—answers the question of *how* the light from the sun, moon, and stars appears to us.

C

THE ARCHAEOLOGICAL EVIDENCE

Metallurgists, blacksmiths, and cooks all used bellows in their operations.[76] And blacksmithing at, or connected with, the archaic temple building sites was routine. The smelting and melting of metal required extraordinarily high temperatures, and these were achieved through the introduction of forced air (which produced a high concentration of oxygen). For the cook's fire, bellows allowed some regulation of temperature and assisted in the starting of the fire. While the bellows technology is the same, the industrial applications require higher temperatures and thus more air. An increase in air volume is achieved with bigger air bladders and a system of two bladders operated alternately. The following is a representative overview of the forms and uses of bellows in Greece, but it is not intended to be an exhaustive one; the main purpose is to place ourselves in Anaximander's world and consider what he and his compatriots would have seen.

C.1

SMELTING

Naturally occurring metal ores contain many contaminants and must be refined before they have a manufacturing use. Archaeological and ethnographic studies have established the smelting process.[77] The furnace shape was probably a cylinder, and was hand formed for only one use out of sundried, local clay. The fuel for the furnace was probably charcoal, which burns at a higher temperature than dry wood. Crushed ore is placed in the bottom of a furnace. There may have been a flux introduced to aid in the separation of ore from its matrix. The top of the cylinder was likely open to allow for a natural draft, but this was supplemented by air from bellows.

Figure 4.1 shows photographs of a preindustrial smelting procedure used in Africa in 1931. The native metallurgists build a bottle-shaped, temporary furnace of local clay, three feet high and with a diameter of one foot, with an opening of around eighteen inches at front. Iron ore is smelt over charcoal, and two blast-pipes attached to bellows increase the flow of air. The blast-pipes are made of unbaked local clay. The narrow end of the blast-pipe extends into the fire. The wider diameter end projects outside the furnace and bellows are attached here. Smelting can take days, but, once the process is completed, the furnace is dismantled and the lump of semirefined ore removed. This ore will require further refining before being manufactured.

C.2

MELTING AND FORGING

Further refining and processing of ore took place in the blacksmith's shop. Two types of metal processing also occurred there: hammer forging of lumps

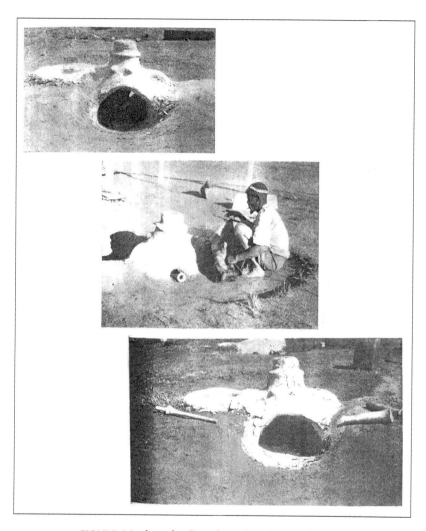

FIGURE 4.1a, b, and c. Preindustrial smelting technique.

of heated ore and casting. Both required extreme temperatures, particularly casting, and these temperatures were reached with the assistance of bellows. In archaic and classical art, the god Hephaistos is shown with the attributes of his trade, that of the blacksmith. These attributes often include tongs (used to hold heated metal on the anvil), a hammer, an axe (a reference to the enormous amount of fuel required for the blacksmith's furnace), and bellows. Not all of the attributes appear in every depiction, and the bellows are the least frequently shown, probably because the painters found their amorphous form difficult to express in the two-dimensional medium. It is important to note that Hephaistos is a blacksmith, not a smelter. He is creating the final product in his workshop. His products are keystones of civilization and his activity is indispensable to the polis. The smelter, on the other hand, is more likely to work at an extra-urban site.[78]

C.3

TWO FORMS OF BELLOWS

There are two possible bellows constructions. The first is made of skin bags; the second of a pot covered with skin. Figure 4.1b depicts the use of a double skin sack bellows by a native African metallurgist. Bellows made of skin bags are often double—that is, there are two skin bags, each ending in a narrow nozzle. In the African analogy, the skins are typically goat. This seems to be true of antiquity as well. Both goatskin bag nozzles are inserted into the blast-pipe, so that each blast-pipe is associated with two air bladders. The double bladder construction allows a worker to maintain a steady flow of air by compressing one by pushing down while simultaneously filling the other by lifting upwards. There are two air intake possibilities for skin sack bellows. Either the goatskin sack is sealed so that the only outlet/inlet is the nozzle, or the sack is open at the top and must be held shut when compressing. In the former case, the outlet/inlet nozzle cannot be inserted directly into the fire, for the air from the fire would fill the sack upon inflation. Instead, the nozzles are inserted loosely into the wide diameter mouth of the blast-pipe so that the intake air comes from the atmosphere outside of the furnace (see figure 4.12 which features pot bellows, not skin bags).[79] The latter alternative is for the bag to fill from an opening at the top of the sack. This is the method shown in figure 4.1. As described in Rickard and depicted in Attic vase painting, the skin bags have slits stiffened with strips of wood or bamboo. To inflate the bag, the slit is opened; to discharge the air, the slits are squeezed shut and the bag compressed. The narrow slits with their frames could easily be manipulated open and shut by each hand.

These modern analogies find confirmation in archaic depictions of Hephaistos and blacksmiths. Figures 4.2 and 4.3 show Hephaistos working the bellows in the context of a gigantomachy. Figure 4.2 is a black-figure kantharos from the Acropolis, circa 550 BCE.[80] This fragmentary representation of a gigantomachy, the battle between the Olympian gods and the giants, preserves a portion of the figure of Hephaistos and a fragmentary identifying inscription to the left of his lower leg. The only preserved attribute of Hephaistos here is the bellows. He is shown operating a two-bag skin bellows.

The large size of the bladders suggests the skin of an entire animal, and indeed details confirm this. The stippling of the surface indicates the hairy pelt of the animal, probably goat. The fur indicates that the skins are not

FIGURE 4.2. Black-figure kantharos circa 550 BCE depicting Hephaistos with (the double bellows the bellows are lower right).

tanned smoothly on the exterior.[81] One leg, tied shut, can be seen projecting from the right skin bag. Three of the legs would have been tied up, with the fourth attached to the nozzle. Hephaistos is in the process of depressing one bag with his right hand and inflating the other with his left. The right bag is better preserved, and on it we see the slit opening at top reinforced with thin sticks. He has clutched these sticks together and is pressing downwards to expel air out a nozzle at the bottom. The nozzle for this right bag is not preserved, but the nozzle for the left bag is partially preserved. It is a cylinder of decreasing diameter attached to one leg of the skin bag. The nozzle is probably made of clay.

Figure 4.3 shows another gigantomachy, this one from the Siphnian Treasury at Delphi (circa 525 BCE). Hephaistos is again shown with the attribute of bellows with the addition of tongs here. The depiction is very similar to that of figure 4.1b. Hephaistos works a double skin bag bellows by depressing the left bladder and inflating the right. The preservation of the relief sculpture obscures some of the technical aspects of the bellows, but the

FIGURE 4.3. Gigantomachy from the Siphnian Treasury at Delphi circa 525 BCE depicting Hephaistos working the double bellows. Reconstruction drawing of the scene, after Moore.

general form is identical to those in figure 4.1b.[82] The skins are clearly animal bodies minus the head; a curly tail can be seen at the far left. Again the skin bag is slit at top and held open and shut with reinforcing sticks. The nozzles of the bellows are not shown, probably due to the deterioration of the sculpture's surface, but a suggestion of the nozzle is present in the extended leg of the skin.

On this red-figure cup by Douris, circa 475 BCE, Dionysos and his entourage escort Hephaistos back to Olympus in a raucous procession (fig. 4.4a).[83] Dionysos leads Hephaistos by the wrist. Hephaistos wears the skullcap of blacksmiths and carries an axe. A satyr follows him bearing bellows and a hammer.

The bellows here are not in use, so the satyr grabs them by the top slit sticks and allows the nozzles to drag behind. The bag is depicted slack, but very hairy. A line drawing of the form, figure 4.4b, indicates that the skin is anatomically correct. This bellows has only one skin bag, but with two nozzles, one for each rear leg. There is only one slit opening shown. Unlike the

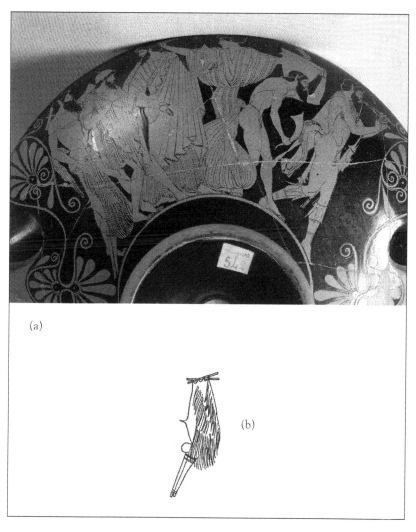

(a)

(b)

FIGURE 4.4a and b. Red-figure cup by Douris circa 475 BCE. (a) Dionysos and entourage lead Hephaistos wearing the skullcap of a blacksmith and carrying an axe. (b) Single bellows with two nozzles, after Gempler.

previous examples, this bellows could not provide a continuous flow of air, but would have to be reinflated (in the same manner as the double skin bag examples), causing a halt in the flow of expelled air.

Figure 4.5 shows a similar single skin bag bellows in a scene with Hephaistos from a column krater by the Harrow Painter.[84] The god is at work at his forge, and he is assisted by two satyrs. The satyrs are references to the Dionysiac return of Hephaistos, and the phallic forge is further decorated with a running satyr. Hephaistos sits on a low stool, and he wears the workers' skullcap. He is about to bring his hammer down on a piece of metal that he holds by tongs in his left hand on an anvil. The satyr nearest the forge holds an axe and gestures to the second satyr for the bellows. The second satyr proffers the bellows.

The bellows are nearly identical to those on the Douris return of Hephaistos (fig. 4.4a). There are clearly two nozzles, but only one skin bag. The nozzles are attached to the back legs; this time the skin is not shown with distinct testicles and penis. Two sticks close the slit opening where the animal's head once was. The skin is painted with a light wash to suggest a furry texture just as in figures 4.2 and 4.4. The implication is that the bellows will be used to stoke the fire of the forge.

FIGURE 4.5. Column krater by the Harrow Painter. Hephaistos works at his forge, assisted by two satyrs. Single bag bellows with two nozzles.

Figures 4.6, 4.7, and the detail in 4.8 all show a scene from a bronze sculptor's workshop. This cup, known as the Berlin Foundry cup, by the Foundry Painter, depicts different end stages of the production of a bronze sculpture. On side A, figure 4.6, two workers are finishing the surface—polishing and smoothing joins—of a supersized sculpture of a hoplite. Two full-sized men look on. Their role is difficult to determine. They may be the workshop owners, patrons, or curious visitors. The positions of the two draped men are typical in red figure for subsidiary figures, and their type appear in gymnasium scenes and often in a group on side B of vases.

On side B, another statue is receiving its final assembly, which is shown in figure 4.7. A worker attaches the hand to a sculpture at right; its head lies on the ground to be attached next. At left is the forge, and a worker wearing a skullcap heats a long, hooked rod. This may be for the rivets needed to assemble the sculpture. Another worker leans on his hammer and waits for the heating of the rod. The forge obscures most of a fourth worker with a rare frontal face; details appear in figure 4.8. He is working the bellows to heat the fire. His position opposite the furnace opening indicates a second opening for the bellows. This placement puts the bellows operator safely out the way of the blacksmith. The bellows operator is shown with his elbow up and out, and in front of him is the slack sack of the bellows. It is impossible to

FIGURE 4.6. Berlin Foundry cup depicting end stages in the production of a bronze figure, side A.

FIGURE 4.7. Berlin Foundry cup depicting end stages in the production of a bronze figure, side B.

FIGURE 4.8. Berlin Foundry cup depicting end stages in the production of a bronze figure. Blacksmith wears skullcap and workman, with rare frontal face, works the bellows.

tell if this is a single sack or double sack bellows. Again, it is difficult to determine the nature of the bellows, but they are shown as an integral aspect of the forge.

The only indestructible, thus archaeologically preserved, element of the skin sack bellows is the nozzle.[85] Nozzles survive from many archaic and classical sites in Greece and around the Mediterranean, including the Athenian agora. Figure 4.9, showing a mid-sixth-century BCE nozzle from a foundry near

FIGURE 4.9. Nozzle of a bellows, mid-sixth century BCE from the Athenian agora.

the Athenian agora, is an example of this type.[86] The nozzles are handmade of clay, probably compressed around a stick to create the central tube. The clay is either fired at a low temperature or simply sun-dried. Some of the surviving nozzles do preserve vitrified areas and patches of metal adhering to the surface.[87] This is evidence of the nozzle's close proximity to the forge fire and melting metal.

In the Roman period, Athenian blacksmiths reused the narrow necks of amphorai as nozzles, as seen in figure 4.10.[88] The shape of contemporary amphorai echoed the traditional nozzle form, and suited the blacksmiths' purpose. That the nozzles were never specially made of fired clay suggests that they had to be replaced frequently, and the investment in more substantial materials was not worthwhile. The reuse of amphorae parts, presumably discarded from broken vessels, affirms the temporary nature of the nozzles. This limited lifetime of the nozzles is probably a result of the high heat and pressure from the bellows.

106

FIGURE 4.10. In the Roman period, Athenian blacksmiths reused the narrow necks of amphorai as nozzles.

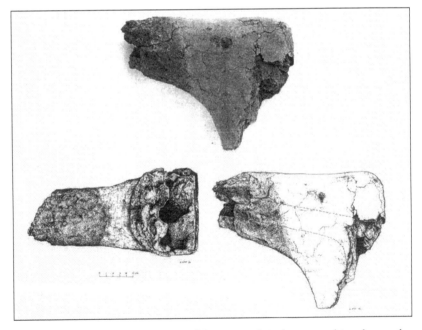

FIGURE 4.11. Bellows nozzle preserved from an archaic bronze working shop under the later "Workshop of Pheidias" at Olympia.

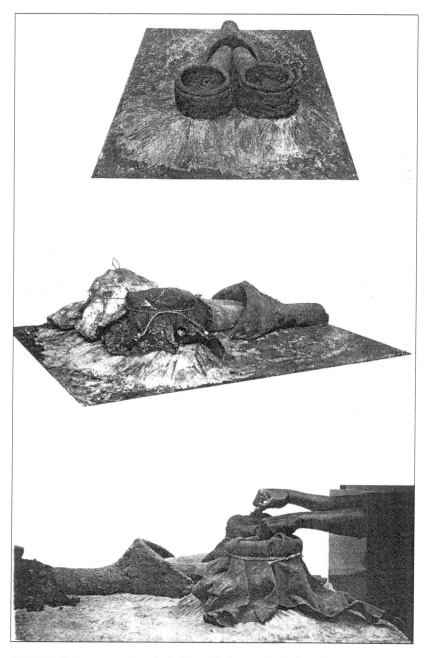

FIGURE 4.12. Reconstruction of a double pot bellows made of clay and covered with skin.

FIGURE 4.13. Nozzle of the bellows from Chrysokamino, Crete, dating to late Neolithic or early Minoan times.

Another type of clay skin-sack bellows nozzle is preserved from an archaic bronze working shop under the later "Workshop of Pheidias" at Olympia (fig. 4.11).[89] The preserved portion of the nozzle features a right angle. The skin sack attached to one end, and the other extended into the furnace. The right angle permits greater control of the direction of the air flow, and may belong to a fixed bellows installed permanently at the furnace (cf. the portability of the bellows shown in figures 4.1, 4.4, and 4.5).

Bellows can also be made from pots covered with an animal skin. These "pot bellows" are known from ethnographic sources and archaeological evidence. Figure 4.12 shows a reproduction of a pot bellows.[90] Two clay cylinders, open at the top and with projecting nozzles, form the body of the bellows. Skins are tied over the tops of open cylinders. In the center illustration, the skin has a loop to facilitate lifting and depressing of the skin.[91] The nozzle of each pot bellows terminates in a large diameter blast-pipe. The nozzles are not sealed and do not form a tight juncture with the blast-pipe, so that air is taken in and expelled from this same aperture, certainly at a loss of efficiency. The two skin-covered pots would be alternately depressed and inflated to maintain a constant stream of air.

Excavations of a late Neolithic–early Minoan copper processing site at Chrysokamino, Crete, have revealed numerous examples of pot bellows.[92]

FIGURE 4.14. Reconstruction of the use of the bellows bladder from Chrysokamino, Crete, dating to late Neolithic or early Minoan times.

The fragmentary fired clay bodies have a large diameter cylinder (28–48 cm), along with a nozzle projecting from the wall (see figures 4.13, 4.14). Unlike the pot bellows shown in figure 4.12, the bellows in figure 4.13 has only a small opening on top, suggesting the bellows bladder arrangement shown in the reconstruction, figure 4.14. The Chrysokamino pot bellows do not have bottoms because the bottom edge of the cylinder would have been buried in soft mud to form a sealed chamber.[93] A blast-pipe was attached to the nozzle. A skin sack was attached to the body of the bellows, and was filled by pulling upwards on a rope at the center of the bladder. It is unclear from the excavation report how air intake for inflation of the bladder worked, nor are the advantages of pot bellows over skin sack bellows readily apparent.

C.3

EPILOGUE

Anaximander's fiery wheels of sun, moon, and stars radiate light through an "opening" in the wheels. Anaximander's hypothesis that the sun, moon, and stars are "wheels" offered a way to explain *why* they do not fall, as would be expected if they were solid bodies. The fiery wheels rotate, going under the earth, and, in the process of explaining why *they* do not fall, Anaximander created another problem of explaining why the heavy earth, held up by nothing, itself does not fall. The answer Anaximander proposed to this conundrum we have already considered in his doctrine of *homoiosis*.

In the usual scholarly recitations of Anaximander's cosmology, the mystery of why we do not actually see "wheels" is answered by supplying that *aêr* or mist surrounds these wheels, produced by evaporation from the cold and wet earth when exposed to the fire. Thus, the compressed air occludes the fire itself and makes it invisible, except at those openings. This cosmic picture, however, rather than solving the mystery instead leads directly to two conundra: The first is to explain how the wheel of fire that is the most distant "sun"—with its fire appearing at its "mouth"—is concealed by the mist air and yet can shine through the occluding mist surrounding the moon and star wheels. If the concealing mist around the moon and star wheels is sufficient to conceal those wheels of "fire," how can the sun shine through, as indeed it does? The second conundrum is to grasp how the moist air can remain so close to this fire without burning off. Let us turn first to consider how the fire from the sun wheel reaches us, and how it penetrates the mist air around the moon and stars.

We do not see the "wheel" of the moon or the "wheel" of the stars because they are concealed by evaporated moist air. But the sunlight reaches us, and so it must be able to shine through the concealing mist that successfully hides the moon and star wheels. How should we understand this? Bodnar addressed just this issue, but his ingenious proposal brings with it a connected problem. Bodnar's explanation follows in the tradition of Kahn, diverging only slightly.[94] His solution is that the amount of concealing mist around the moon and star wheels must be less dense—that is why the more fiery sunlight can shine through. Here is what is at stake in resolving this problem.

There have been two strategies in the secondary literature to account for the unusual order Anaximander proposes of sun, moon, and stars (from furthest to closest). The analytic approaches reach for a "coherent" explanation (i.e., by their approach they term "rational") of this unusual order. They ask, in effect, what line of reasoning coherently explains this unusual order, that is unusual even for the ancient Greeks. The nonanalytic approaches following Eisler and then Burkert and then West suggest that Anaximander adopted the Zoroastrian account transmitted to him and his compatriots when the *magoi* were dispersed from Persepolis and wandered down and along the west coast of Asia Minor. According to the *Zend-Avesta*, Ahura Mazda travels first to the sun, then the moon, and finally the stars before entering the fiery hearth or Hestia in our own homes. West observed that there is no hint in the meteorological tradition of the Greeks to defend this cosmic order of sun, moon, and stars, and the transmission from the dispersed Zoroastrians is hard to discount. Kahn had argued, earlier, that the Greeks held the reasonable view that fire tended upwards; the more fire, the further away would be the circular wheel. Since the sun is clearly the hottest fire, it is reasoned that it must be furthest, the stars closest as a matter of simple observation since they are least bright, and the moon intermediate between them. Bodnar asks us to consider *how* it is that the sun's fiery wheel is invisible, unable to shine through its encompassing *aêr* mist, and yet the fire appearing at its "aperture" (Bodnar's rendition) nevertheless can shine through to reach us. Bodnar proposes that the mist envelope decreases in density around the moon and then the stars—because they are less hot and consequently produce less evaporation. If the density of the enveloping mist around the moon was the same as around the sun, the sunlight at the opening of its wheel could never shine through to reach us. This proposal seems plausible to me, but it brings with it another complication. The underlying problem is to be clear about the diameter of each wheel; they all seem to be one module in diameter. But what keeps them to this diameter? It seems that it is only the quantity of fire plus compressed moist air. Now if this is right, and if the amount of fire is less in the moon wheel, and even less in the star wheel, as it appears, and, correspondingly, the amount of concealing *aêr* is less and less, then the diameter of the wheels should also be different from the sun wheel and from each other. But, if this is the case, then the cosmic numbers must be wrong. Let us reflect further on this matter.

We have argued that, in addition to the testimony that the sun is the size of the earth, an appeal to the application of architectural techniques to the cosmic architecture suggests that each wheel is one module in diameter. Then, the cosmic numbers make sense based on the doxographical reports, the archaic formulas of 9 and 9 + 1 sung by Homer and Hesiod, and also by appeal to the archaeologist's argument for "technological style."[95] The module is for Anaximander, as it is for the archaic architects, column diameter, the earthly measure in terms of which Anaximander reckons the whole cosmos. If the density of the *aêr* mist decreases around the moon and stars, then it would seem that the fire would not occupy the same modular space. For what is stopping the mist from compressing closer, if not the amount of fire in each wheel? If the density of surrounding mist decreases in proportion to the lesser amounts of fire in the moon and star wheels, the cosmic numbers should have been different. Then each wheel would not have been "one module" in diameter. So, our conundrum is that if we are to accept Bodnar's reasonable proposal that the more fiery sunlight permeates the compressed *aêr* surrounding the moon and star wheels because the moon and star wheels are surrounded by less dense moist air, then the moon and star wheels should have a smaller diameter and the cosmic numbers must be different. Since we are told in the doxographical reports that Anaximander maintained both cosmic symmetry and geometrical proportions, and we have the numbers 27 and

28 for the sun wheel distance and 19 for the moon, the "changing-density of the compressed mist thesis" offers to resolve one problem at the cost of creating another. But perhaps there is another way to think about this. Hippolytus informs us that the heavenly fire appears at "breathing holes" that he describes further as "certain pipe-like passages" (*ekpnoas d' huparchai, porous tinas aulôdeis, kath' ous phainetai ta astra*).[96] Could it be that Anaximander imagined the heavenly wheels as some kind of plastic medium, like stovepipes? While the lack of evidence for details must place such an hypothesis in doubt, a pipe-like passage whose diameter was not a function of the quantity of fire within it would allow us to retain both the cosmic numbers 9/10, 18/19, 27/28 and also the idea that the density of evaporated mist concealing the wheels diminished in proportion to their diminished fire. On this interpretation the amount of fire in each wheel decreases from sun, to moon, to stars, but not the modular size of the wheel itself that remains one column-drum diameter. The size of the wheel is not a function of the compressing mist, though the mist surrounds the fiery wheel.

A further inquiry leads to another conundrum: *How* is it that the sun wheel causes evaporation from the cold and moist earth, and so is occluded by mist, that is yet able to envelope and compress the fire *without dissipating?* The sun is very hot, as any visitor to Greece and Turkey in summertime would know firsthand; the whole wheel, at the very beginning when it was unoccluded, must have been immeasurably hotter. So why does the mist not simply burn away? This question might have led Bodnar to see that *aêr* mist "feeds" the fire in cosmic, cyclical transformation, that rather than fire and moisture being simple opposites, they are inextricably connected in cosmic cycles. Moreover, the mechanism of the bellows that is proposed to explain how the fire radiates from the wheel is really part of that mechanism of cosmic breathing through the "blowhole" (more than just an aperture) that Anaximander identifies, as we can see from the art historical evidence provided by the archaeologists of Hephaistos inflating and deflating an animal skin/bellows.

If the fiery wheels produce so much heat to cause the water on the earth to evaporate upwards—the heated moist air producing winds and other meteorological phenomena that cause the sun's "turnings" (i.e., solstices)—how could the moist air remain so close to the fire without dissipating? The answer, it seems, is that for Anaximander, like Anaximenes, in the great cycles of transformation of one material form into another—water into *aêr* into fire into *aêr* into rain into earth and so on, since each material form is only an *appearance of the other*—water evaporates, that is, *exhales as aêr*, and the fire *inhales* this moist air that "nourishes" (τρέφεται) it, and in turn *exhales* the fire as a scorching air or wind. The transformation of one material form into another can be described appropriately as "inhalation" and "exhalation" for Anaximander, and this doctrine is explicitly affirmed in the surviving doxographical reports attributed to Anaximenes, who shares a similar vision on this issue. On the authority of Hippolytus,[97] Anaximenes maintained that *gegonenai de ta astra ek gês dia to tên ikmada ek tautês anistasthai, hês araioumenês to pur ginesthai, ek de tou de tou meteôprizomenou tous asteras sunistasthai:*[98] "The heavenly bodies have come into being from earth through *exhalation* arising from it; when the *exhalation* is rarefied, fire comes into being, and from fire raised on high the stars are composed."[99] To speak in terms of "exhalation" and "inhalation" is to describe the cosmic process of condensation and rarefaction that generates one elemental form from another on analogy with human life. In Anaximenes, perhaps, it is immediately more apparent because the very same stuff we breathe, that gives us life—*aêr*—for without it, we die—is also *pneuma*, the soul or life breath.

The idea that moisture feeds fire in cosmic transformation has much support. Aetius proposes a reason why Thales might have chosen water as his source: "[Thales may have chosen water because] the very fire of the sun and the heavenly bodies is fed (*trephetai*) by exhalations of water, as is the world itself."[100] The theme that heavenly bodies receive nourishment from moisture is also found in Anaximenes.[101] Despite some complexities in the surviving reports, Xenophanes, too; but at all events, the stars are fiery clouds, the moon is a felted cloud, and the sun is fed my moist exhalations.[102] That this point of view—that the heavenly bodies were fed and sustained by moist exhalations—was widely known and influential can be surmised by a text from the late fifth century preserved in the Hippocratic library: "The path of the sun, moon, and stars is through the air [*pneuma*]. For air is fuel [*trophê*] of fire, and fire deprived of air would not be able to survive. So thin air supports the everlasting life of the sun."[103]

Anaximander described the "opening" in a fiery wheel as a *stomion* or an *ekpnoê*; the opening is a mouth or blowhole. It is surprising that in Couprie's thoughtful essay he nowhere mentions, acknowledges, or considers these crucial terms from the doxographical reports, and thus fails to consider an organic interpretation of this cosmic mechanism. When we take the hylozoistic clue, we find ourselves viewing, metaphorically speaking, cosmic respiration, a part of the cycle of elemental transformation. In the absence of more detail, of course, we are left at a speculative threshold, to understand more precisely the specific mechanical intricacies of this cosmic breathing. But the archaeologist has been able to supply us with images that inspired such cosmic imagination among the archaic Greeks.

In the image of Hephaistos working his bellows, an anthropomorphic cosmic technology, we are clearly not conjuring the nineteenth-century image of a bellows, more like a two-armed fan, one held with each hand and pressed in and pulled out like an accordian. While the Egyptians had bellows worked by the feet to stoke their smelters, the image of Hephaistos supplies the raising—inhaling—of one great animal (skin) while exhaling the other, and producing a thunderous blast. It is impossible to escape the image of an animal coming alive through its inhalation, its whole body filling with air and exhaling in the same connected process, the moist air *transforming* into fire, a fiery scorching wind. When we reflect on this imagery of a bellows that is suggested by the *organic* interpretation of *prêstêros aulos*, we can see more clearly how this unusual expression is fitting for the Milesian hylozoist. For the archaic audience, the expression conjures something much more than a *physa*; instead it conjures a blowing and burning in the cosmic exhalation, part of the transformation of the nourishing moist air to a replenished fire. The expression *prêstêros aulos*, then, supplies an answer to how the fiery wheels radiate light, through a mechanism that is part of the living cosmos.

Thus, the objection that the bellows requires two holes and our evidence for Anaximander's cosmic wheels tells of only one is a *petitio principii*. The "objection" that it seems "odd" that so many bellows mechanisms would be required to account for all the heavenly lights is hardly an objection at all. The opening through which the fire appears is a single hole like a "nozzle"; how precisely the wheels "breathe" is not supplied, but the clues of "mouth" and "blowhole" indicate that the cosmic wheels are breathing and the light that appears to us is a kind of cosmic exhalation. The "opening" in the wheels is a "breathing-out-hole,"[104] and Diels' scholarly insight to render *hôsper dia prêstêros aulos* "as through the nozzle of a pair of bellows" rightly preserves the imagery of that breathing mechanism in a cosmos that is *alive*.

FIVE

Anaximander's Cosmic Picture

The Heavenly "Circle-Wheels" and the Axis Mundi

A

THE DOXOGRAPHICAL REPORTS

Ἀναξίμανδρος [sc. τὸν ἥλιον φησι] κύκλον . . . ἁμαρτείῳ τροχῷ παραπλήσιον, τὴν ἀψῖδα ἔχοντα κοίλην, πλήρη πυρός.
—Aetius I, 20.1, D-K 12A21

Anaximander [says the sun] is a circle . . . resembling a chariot wheel, having a hollow rim, full of fire.

Ἀναξίμανδρος [τὰ ἄστρα εἶναι] πιλήματα ἀέρος τροχοειδῆ, πυρὸς ἔμπλεα.
—Aetius II, 13.7, D-K 12A18

According to Anaximander, the heavenly bodies are wheel-like compressed masses of air filled with fire.

B

THE SCHOLARLY DEBATES OVER
THE TEXT AND ITS INTERPRETATION

THE SCHOLARLY LITERATURE is in broad consensus about Anaximander's cosmic wheels. Anaximander imagined the sun, moon, and stars to be circular rings of fire, encased in mist. He likened them to hollow wheels. The fiery wheels formed when an original surrounding fire broke off into separate rings.

By encasing these wheels in mist, Anaximander provides a way for us to understand why we do not see the "wheel-like" shapes themselves. The proposal that they are circular bodies requires us to see that Anaximander dared to imagine that the sun, moon, and stars go *under* the earth; if this is the case, then the earth cannot be supported from below by any foundation that customarily could be regarded as material. About these claims, there has been general consensus and assent in the scholarly literature.[1]

There are two key points in identifying Anaximander's picture of sun, moon, and stars: (1) their shape is circular; and (2) they are likened to wheels. In stating the matter this way, two different points are made, namely, that while their geometrical form is a circle, their shape and mechanism are illuminated by analogy to the technology of wheels. What has been missing in the debates among scholars is an attempt to get clearer about just what kind of "wheel" Anaximander imagined. What has been missing is a thorough investigation of ancient wheels and wheel making. Did Anaximander actually see any wheels that had hollow rims that might have been the source of his analogous reasoning or metaphorical projection? Had he watched craftsmen making wheels, would he have discovered some clues in wheel-making technologies that might shed more light on the cosmic wheels? And when we review possible candidates for the hollow-rimmed wheel that Anaximander might have seen, are any more clues provided about the overall cosmic picture?

Accordingly, there are four objectives of this presentation of archaeological resources:

1. Scholars of ancient philosophy have not had a resource for the archaeological evidence of wheel making and its technologies. Not only for Anaximander, but also, for example, for the chariot Parmenides mentions at the opening of his poem, when "the axle at the center of the wheel was shrilling forth the bright sound of a musical pipe"[2]—is this the sound of a fixed axle or rotating axle? In general, what specific technologies and mechanisms come to mind for the archaic Greeks when Anaximander likens the sun to a wheel? A central purpose of this chapter is to provide an exploration of ancient wheel making and its technologies, a resource for future work in ancient philosophy.

2. A review of the ancient evidence, it will be argued, supplies an example of a wheeled vehicle with a "hollow rim"; it is a vehicle for transport of architraves for a monumental temple, credited to the architect of the archaic Artemision, Metagenes, and almost certainly featured in his prose book. This is a conclusion we shall reach by the end of the survey.

3. An exploration of wheel making supplies technologies for tripartite wheel construction, among other techniques. Anaximander is credited with making the first Greek map of the inhabited world, a tripartite map. Since

the *pinax* was circular like a wheel, we shall explore whether techniques for making ancient wheels hold any clues for the divisions in Anaximander's map. And we shall follow up this discussion when we explore Anaximander's seasonal sundial in the next chapter.

4. An investigation of ancient wheel making shows that a wheel had a fundamental meaning in the context of a connection with an axle. Wheels were not, as practical devices, free-floating geometric shapes but rather utile elements whose success required an interconnection in a system that facilitated the turning of the wheels. When we re-view Anaximander's cosmic wheels in the context of axle construction, will we see that Anaximander had also pictured a cosmic system with a cosmic axle—an *axis mundi*?

To proceed to these objectives, it will be useful to review, briefly, the evidence for heavenly *wheels* in the context of the whole cosmic picture.

On the authority of Aetius, writing in the second century CE, Anaximander's theory of the sun, moon, and stars is partially explained.[3] Anaximander held that the sun is not an isolated body but rather a *kuklos*, a "circle," many times the size of the earth. In describing that geometrical form, Aetius reports that *Anaximandros (ton helion phêsi) kuklon . . . hamarteiô trochô paraplêsion, tên hapsida echonta koilên, plêre puros:*[4] "the sun is a circle . . . it is like a *hamarteiô trochô*[5] with the curved and circular part of the wheel hollow, and it is full of fire."[6] It is central to this chapter to get clearer about the meaning and the reference of the expressions *kuklos*, "circle," and *hamarteiôi trochôi*, "chariot wheel," as well as the imagery that is conjured in the process, and perhaps also the method or process of producing such wheels. When we do, perhaps, we shall see more clearly how Anaximander imagined these cosmic wheels.

In the scholarly literature, there is little or no debate about this testimony, that is, that the sun, moon, and stars are "circular" in shape and that they are likened to wheels with a hollow rim. The wheel shape is explained by some eternal motion by means of which a fire that surrounded the periphery of the cosmos was separated out into fiery wheels; the compressed moist air around each wheel is caused by evaporation from the fiery heat. The fire that *is* each wheel is occluded by that compressed moist air except where it appears at a vent-like opening described sometimes as a mouth (*stomion*) or a blowhole (*ekpnoê*). Hippolytus describes that the fire appears as "breathing holes" that are *porous tinas aulôdeis,*[7] sometimes translated as "pipe-like passages"[8] and other times "tube-like passages."[9] The translations of *hamarteiôi trochôi* are also twofold: some translate (a) "like a chariot wheel with its felloe hollow and full of fire,"[10] while others, such as Guthrie, translate (b) "resembling a cartwheel with the rim hollow and full of fire,"[11] thus calling upon a wheeled vehicle other than a chariot. There is no doubt that the sun, moon, and stars are circular in shape and likened to some kind of wheel, which was pivotal to Anaximander's understanding why the sun, moon, and stars did not fall from the heavens. Was Anaximander imagining a particular kind of wheel? Could he have seen a wheel with a hollow rim that fueled his cosmic imagination?

By some mechanism, not initially clear from the surviving testimony, an original fire at the outermost part of the cosmos somehow separated out or separated off to form these fiery rings.[12] Due to eternal motion,[13] probably vortex-like, the heavenly wheels formed. The reason why we do not see fiery wheels, however, is because the wheels are encased in *aêr*, moist air or mist.[14] The *aêr* is compressed sufficiently to conceal the wheels, yet the *aêr* is invisible to us. Aetius goes on to explain that what we actually see, in the case of the sun, for instance, is the fire that fills the wheel but only at an opening. This opening is a "mouth" (*stomion*), and the fire radiates through that opening just as it does at the nozzle of a bellows, or, following Diels, through the nozzle of a pair of bellows. In another passage, also on the authority of Aetius, Anaximander describes the opening at which the fire appears as *ekpnoê*, a "breathing hole."[15] This is a remarkable description of the heavens for several reasons. If we can accept as reliable at least the main outlines of Aetius' report, Anaximander appealed to technological metaphors to illuminate celestial events. Thus, he was able to project metaphorically, into and onto the heavens, the dynamics of earthly mechanisms that he and his compatriots could observe closely in order to reveal cosmic structures and mechanisms that at best could only be inferred. Additionally, the image of the cosmos that can be derived is a fully organic one; the whole cosmos is alive, and as each heavenly wheel is breathing, the cosmos itself is breathing. The fact that Anaximander's heavenly cosmos is alive through an everlasting breathing of fire, a point that anticipates Heraclitus, often remains unacknowledged. In this chapter, we shall investigate the particular technical image and mechanism of wheels used in transport in the effort to *imagine* more clearly Anaximander's heavenly structures and mechanism. Why and how should we do so?

In *Anaximander and the Architects*, the discussion was set in motion by appeal to two pieces of the puzzle of grasping Anaximander's cosmic imagination. The first was the identification of the shape and size of the earth with a *column drum*,[16] and the second was the recognition that the distances from the earth to the wheels of the stars, moon, and sun were reckoned in *column-drum proportions*. With these two architectural clues in hand, the investigation proceeded to explore the gigantic architectural projects going on, for the first time, in Anaximander's Ionian backyard, and to reflect upon what he could have learned from them had he taken the time to do so. The result was to see that the identification of the size and shape of the earth in terms of column-drum diameter was no throwaway or casual remark. These clues led to the recognition of the techniques that the architects were routinely practicing in the temple sanctuaries in Didyma, Ephesos, and Samos, contemporaneously and slightly earlier. Among the techniques that proved most relevant for Anaximander's speculations, the *theory of proportions* deserves to be highlighted.[17] The architects planned and marked out the dimensions of their great temples in the proportions of a basic module, and the architectural module itself in Ionia was column diameter.[18] Anaximander's reckoning of the size of the cosmos in terms of column-drum proportions, then, revealed not only that he followed the modular technique of architects, but also that he made use of precisely the same module that they did. The investigation helped us to see Anaximander's mentality, most especially his reasoning by metaphor and analogy. He imagined cosmology in architectural terms—the stages and structures of monumental building—thus connecting terrestrial with cosmic architecture. This leap of thought, explaining the structure of the heavens by appeal to techniques in architecture, connecting the microscopic earthly structures to the macroscopic cosmos, was facilitated by means of architectural techniques. Thus, *Anaximander and the Architects* showed that the architectural clues pointed to discussions in his archaic community, which we have now begun to overhear, about the cosmic and symbolic meaning of the column, and the techniques by which the houses of cosmic powers were appropriately calculated. Moreover, the connection demonstrated that Anaximander was able to project metaphorically from terrestrial architecture to celestial architecture, just as he did from terrestrial cartography to celestial cartography.[19] Now, when we take up the "chariot wheel" clue to explore further the cosmic structure and mechanism, we should keep in mind this microscopic-macroscopic projection in Anaximander's thought, for it operates throughout.

When the architectural connection was pursued, we were forced to imagine, perhaps for the first time, Anaximander at the building site as a careful observer. The array of techniques that he, and his contemporaries, would have witnessed in the building activities was explicated. Those techniques included the stages of planning, drawing, and marking out initial dimensions according to the theory of proportions. Next, we considered the other techniques that could have been easily observed such as the quarrying of stones, transporting them to the building site, the steps of preparing the blocks, especially column drums, and installing and finishing them. Through this analysis, a much more complex picture began to develop of the powerful images and technologies that came to assume a formidable stature in Anaximander's mind. The architects through their technologies supplied projects for the archaic community to reflect upon and, by means of architectural prose books, they promoted prose narratives fundamentally rational in character, discussing nature's structure and dynamism. In turn, Anaximander, sharing a comparable spirit that has hardly been appreciated, expanded their projects by means of his own prose book. His new narrative offered to supply a rational, not poetic, explanation of nature's order at great distances. Rather than focus on temple architecture and the house of the cosmic and divine power, Anaximander sought to explain the structure and stages and construction of the house that is the cosmos. The temple was part of an instrument of mystification. Humans were supposed to be unable to penetrate nature's secrets; the temple celebrated the incapacities of mortals. The architects opened a new doorway into nature's patterns, demystifying natural laws. And Anaximander, following this new opening, further rationalized and thus demystified nature.

C

THE ARCHAEOLOGICAL EVIDENCE

Now, we must begin to make another journey back into archaic Greece, but the exact location is still an issue. Perhaps to answer our inquiry into the technology of wheel production, we must venture into the workshops in the agora, or at least the industrial section of Miletos.[20] Or, perhaps, as seems more likely, we must return again to the great building sites of Didyma, Ephesos, or even Samos. For at these sites, no doubt, a variety of vehicles of transport were certainly in use conveying materials and workers. At these sites, work would certainly have been under way to repair broken wheels and axles, and so to aid in the delivery of weighty blocks from the quarry, usually many kilometers from the building site. And, of course, so many technologies would have been displayed there. Thus, our question now is to ask what more Anaximander could have witnessed in the workshop of the wheelwright, chariot maker, or at the building site.

Thanks to Aetius, we have a clue, but no details. When we supply those details, however, a very rich source of techniques and technologies begins to appear. Our starting point is the simple question: What specific images of the wheel did Anaximander have in mind when he suggested that the wheel-like sun, moon, and stars could be communicated by means of the analogy? As we begin to explore what he might have had in mind, we are challenged by two central concerns. The first is to acknowledge that there must have been many wheeled vehicles for which we have no surviving evidence, and these particular wheels, for chariot and transport vehicles, may have been the ones that Anaximander specifically had in mind. The second concern is that when we begin to explore the range of wheeled vehicles with which he could have been familiar, the technology of making the wheels is seen to merit additional attention. For not only is the wheel imagery central, but so also is the technology for its production, for this technology helped Anaximander unravel the cosmic mechanism by which it turned. As we have seen with architecture, so also we shall make the case for wheel production; both architectural techniques and activities at the building site influenced Anaximander's cosmic imagination. At the building sites, there would have been workshops to repair the wheels on vehicles of transport. This means that techniques for wheel making[21] and the integration of them with fixed and moveable axles were present to Anaximander's view, and this means also that techniques for using a forge and bellows would have been routinely on display.

C.I

THE ARCHAEOLOGICAL EVIDENCE
FOR EARLY WHEELED VEHICLES

The seminal work on ancient wheeled vehicles was produced by V. Gordon Childe. In important articles exploring the earliest evidence for "Rotary Motion" and "Wheeled Vehicles," Childe offered an overview of how the principal of rotary motion found its way into the services of the potter and for those in search of transport for themselves and their goods.[22] While there is scant evidence for wheeled vehicles earlier than the third millennium BCE, the evidence is comparatively robust for wheeled vehicles subsequently. Childe traced out the evidence on vases, limestone reliefs, and pictographs from Susa, Assyria, and greater Mesopotamia in general.

The oldest and most widespread of all wheel formations, Childe argued, was the *tripartite* disk (see fig. 5.1), although it was neither the only conceivable form nor theoretically the simplest.[23] The appearance of many examples of this type provides a cogent argument for its wide diffusion. While a wheel made from a single plank of wood would seem to offer the simplest form of construction, the absence of trees large enough to yield a single plank may have been one reason why single plank wheels did not predominate. Even in well-wooded countries, examples of tripartite wheels appear side by side with single-plank ones. Examples from Susa dating to the mid-third millennium such as the one pictured in figure 5.1 were formed by taking three planks of wood encircled by a rim, also apparently made of wood. The exact means of forming the rim cannot be precisely determined from the surviving example. It might have been formed from several segments joined by a kind of mortar, or a single strip bent by heat, or even after soaking in water. The rim was studded with small copper nails that project from the rim in cog-like fashion. The function of the nails was certainly to protect the rim from wear. This device was retained, as the evidence from Assyrian and Achaemenid Persians attests, even after metal tires, surrounding the rim, had been introduced.[24]

In other early examples from Ur, we have evidence that the copper nails were used to fasten leather tires to the tripartite disk, as well as evidence, even at this early date, for the use of metal tires, made of copper. Had Anaximander's tripartite map followed this example of wheel making, the illustration would have been immediately relevant, as seen in figure 5.2.

In figure 5.3, four or six concave bands of copper constituted the segments of a circle. The parts were fitted over and clamped to the rim of the tripartite

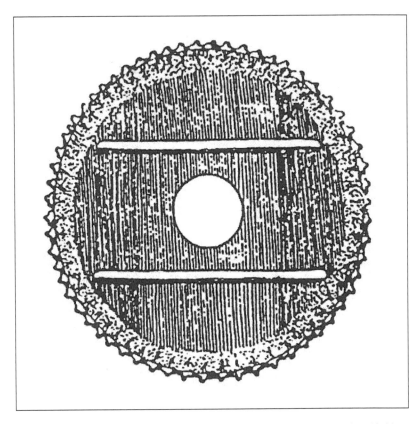

FIGURE 5.1. Nail-studded tripartite wheel from Susa, circa 2500 BCE, after Childe.

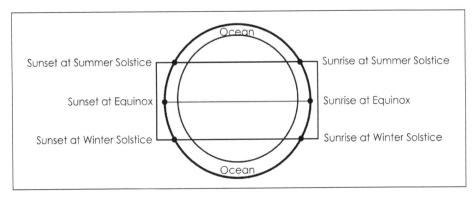

FIGURE 5.2. Anaximander's earth with sun risings and settings on solstices and equinoxes; the rectangle suggests the inhabited earth. (See also fig. 6.4 and discussion, following Aristotle's report in *Meteorologica* II.6.)

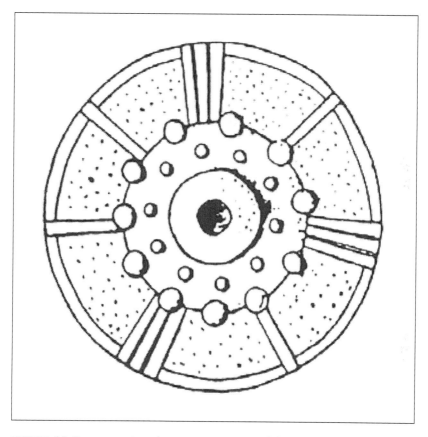

FIGURE 5.3. Representation of a copper tire on a model wheel from Susa, third millennium, after Childe.

disks with the aid of a pair of tongues projecting radially toward the hub from the ends of each tire segment.

While solid wheels are sufficiently strong and efficient for the transport of goods, they proved heavy and clumsy when speed and maneuverability were required. To address these needs, spoked wheels were invented. The spoked wheel quickly replaced the solid wheel on war chariots where pace and control were crucial. However, spoked wheels were more costly than the tripartite ones since they required more labor and skill to produce. What kind of technology was required for the competent construction of spoked wheels?

Spoked-wheel construction required three central ingredients: (1) a hub with an axle hole and sockets for the (2) spokes, four, six, or eight spokes, and (3) a rim or felloe. Each of these parts had to be fashioned separately. The hub is usually carved out of a single block of wood, though there are Egyptian examples that are composite. To function effectively as a wheel, each spoke must be exactly the same length and then trimmed to achieve an exact fit at both the hub and rim. Thus, precise measurement and exacting labor were necessary to produce a serviceable wheel. And just as in any experimental science, the proof of the technique consisted in the deployment of the chariot. Trial and error surely played a prominent role in perfecting the wheelwright's craft. Numerous examples of spoke-wheeled chariots appear in the second millennium from Egypt, Crete, and throughout Mesopotamia. We even have an Egyptian scene in a Theban tomb depicting wheelwrights dating to the New Kingdom (fig. 5.4).

In order to approach more closely the structure of chariot wheels and their production that Anaximander may have witnessed, we must focus on the archaic Greek evidence. In addition, we must consider not only the image of the wheel but also the construction of the axle and the process by which it was produced. How did it function? How does the construction reveal the mechanism by which it worked? We now turn to the evidence for archaic chariot wheels. We will do well to keep in mind, however, that the tripartite wheel long continued to be produced, alongside the spoked wheel, and perhaps this wheel was the focus for Anaximander's reflections. An additional consideration was that Anaximander is credited with making the first Greek map of the inhabited world; he imagined the earth as having a round shape and so the map would have appeared on a wheel-shaped earth. Could the tripartite wheel have seemed especially suitable for a tripartite map?

FIGURE 5.4. Egyptian wheelwrights from a New Kingdom tomb in Thebes, circa 1475 BCE, after Childe.

C.2

The Archaic Greek Wheel and Axle

In the minds of archaic Greeks in general and Anaximander's contemporaries in particular, a wheel was not simply a geometric shape. The wheel was part of a complex system that required the fitting of an axle for its use. This is a key point to remember as we explore further.

Although no actual archaic Greek wheeled vehicles—chariots or carts— survive, visual representations appear in several media: vase painting, relief sculpture, and terracotta and bronze figurines. In *Chariots and Other Wheeled Vehicles in Iron Age Greece*, J. H. Crouwel discusses the surviving artifacts.[25] From the archaic visual evidence, we can discern two types of axle configuration. A fixed axle, around which the wheel rotated, permitted speed and agility for fast-moving vehicles such as the chariot. A rotating axle was more suitable for slow-moving carts with heavy loads. Since the two require different construction techniques, we will examine each in turn.

C.2.A

Fixed Axle: Chariots of the Mainland, Figures 5.5–5.7

The chariot was primarily used in racing and the *apobates* competition as well as in processions, yet its heroic connotations made it an extremely common iconographic element in archaic art.[26] The conventional imagery of vase painting emphasizes the relationship of mortal activities and those of gods and heroes. This relationship is often alluded to through unrealistic conventions such as heroic nudity. The chariot, although a frequent image, refers metaphorically to heroic activity and does not record actual practice. Nonetheless, Crouwel emphasizes the consistency in chariot depictions and points to actual racing chariots as sources for the images and their details.

The fixed axle is mounted beneath the chariot body floor. In profile two-dimensional representations of chariots, the axle end is shown round and occasionally with linchpins securing the wheel, as in figure 5.5.[27]

The axle appears on side A of this black-figure neck amphora; a beardless youth and a bearded man each stand in a chariot pulled by two horses. The youth and his team are in the foreground and the man and his team are behind. It is likely that the pair are either preparing for a race or participating in a procession.

The chariots are nearly identical in representation. The wheels have four bipartite spokes, wedge-shaped supports for the spokes at the felloe, a composite felloe, and a clearly depicted linchpin. The axle end is circular, and a vertical, keyhole-shaped element at its center indicates the linchpin used to hold the wheel onto the axle. The painter has indicated that the spokes are

FIGURE 5.5. Black-figure neck amphora with two chariots, Piraeus Painter, after Crouwel.

made of two separate pieces, and at the nave the spokes feature clear criss-crossed lashings. These bipartite spokes are lashed together with rawhide thongs near the nave and further strengthened by the wedge-shaped supports at the felloe. The painter has also shown two concentric circles for the felloe. The composite felloe may be either two pieces of bent wood or a wooden felloe covered with rawhide tire or even an iron ring tire.[28] Whatever the construction, the painter has perceived two separate portions of the felloe and rendered them as concentric circles of equal width.

In frontal two-dimensional representations, the axle is seen to project beyond the width of the chariot body, as in figure 5.6.[29] The images are often awkward as the painters struggled with the limitations of the profile conventions of vase painting. Nevertheless, frontal depictions do provide more details of the fixed axle chariot unseen in profile views. In this depiction, the four-horse team stands at rest in front of the chariot. A charioteer is upright in the chariot body and holds the reins. The chariot car is proportional to others in vase depictions, but the frontal view reveals the surprising length of the axle and width of the wheelbase. The details are obscured by the strict profile view of figure 5.6. The long axle and wide wheelbase would give the chariot stability in turns at high speed. There is no indication of a separate tire on the felloe.

The long axle clearly passes under the bottom of the chariot body. The painter has shown the naves slipped over the thinner axle and secured with a linchpin at the axle ends. Each wheel has a distinctively rendered nave. The painter's nave is of the same general form as the one depicted in figure

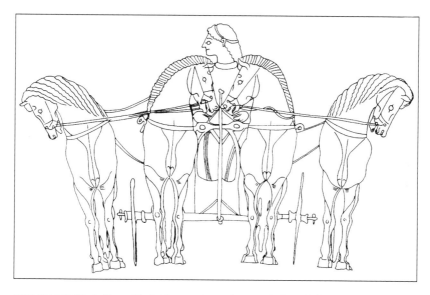

FIGURE 5.6. Black-figure Chalcidean neck amphora with frontal chariot, after Crouwel.

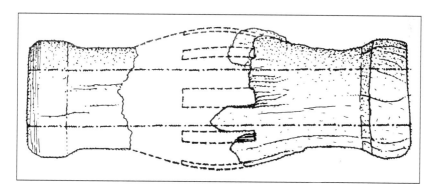

FIGURE 5.7. Fragmentary wheel nave from Olympia, after Crouwel.

5.7, the nave from Olympia. The nave fits as a cylindrical sleeve over the axle that extends beyond the nave. It is possible that the axle diameter is reduced at the nave, for the diameter of the axle projection with linchpin is shown smaller than the diameter of the axle under the body; however, there is danger in reading vase paintings too literally as snapshots of reality. Accuracy does not appear to be the goal of the painter.

Crouwel states that the advantage of a long axle is the ability to accommodate a long wheel nave, which was necessary to prevent the wheels from wobbling while at the same time giving lateral stability in navigating turns. The nave must fit the axle snuggly in order to prevent road dirt from creating dangerous friction. Both units must have been oiled or greased to further reduce friction.

Both the wheel and the axle are made of wood. The wheel is constructed of a central nave with insertion slots for the spokes. A surviving fragmentary wheel nave from Olympia, carved of a single piece of elm, has a central hole for the axle and oval mortise slots for eight radiating spokes (fig. 5.7).[30]

Although eight-spoked wheels are rare in archaic and classical chariot depictions, this fragmentary wheel nave with eight spoke mortises confirms that such wheels were known in mainland Greece. This fragmentary nave made from a single piece of elm is preserved because it was tossed into a water-filled well. The cylindrical form of the nave is thicker at the center where eight mortises received the spoke tenons. The estimated length is approximately 0.38–0.42 m. Crouwel states that ancient naves tended to range from around 0.35 to 0.45 m. The spoke mortises are rectangular with restored dimensions around 0.077 by 0.02 m. The spokes were of single-piece construction. The eight-spoked wheel is more common in eastern Greece where vehicle construction followed styles of the Near East more closely.

The inner diameter of this nave is around 0.06 m, and would have accommodated a fixed wooden axle of slightly smaller diameter. The axle end would have projected beyond the nave, and the wheel would have been secured with linchpins. Oil or grease would have reduced friction of the nave as its wooden surface rubbed against that of the fixed axle.

It is impossible to know if this wheel had a tire, but an iron tire "sweated on"—that is, applied when red hot and then cooled and contracted—would add contracting force to hold the felloe, spokes, and nave together.[31]

Eight spokes is an unusual number in surviving mainland archaic Greek representations, but not unknown. Typically, chariots are shown with four spokes and occasionally with six.[32] Although the Olympia nave featured simple, one-piece spokes, two-piece spokes are more common in archaic depictions.[33] These are clearly marked by rawhide lashings used to bind the two halves of the spoke. The rim of the wheel, the felloe, is also made of bent wood. In the case of simple, mortised spokes, the wheels probably bore iron tires attached not with pins but by sweating them on in order to provide the force necessary to retain the structural integrity of the wheel. The alternative, bipartite spokes have the advantage of distributing the forces throughout the wheel at all times and thus did not necessarily need a constricting "tire." These may have had a rawhide tire, also shrunk on, to offer some extra constriction in addition to protecting the edge of the felloe. However, some depictions clearly show bipartite spokes and a "composite" felloe—that is, an outer rim with two separate portions. It is possible that the outer ring is an iron tire sweated on as previously described, but the only surviving evidence for iron tires in archaic Greece is associated not with chariots but with carts.[34]

C.2.B

FIXED AXLE: CHARIOTS OF EASTERN GREECE
(CROUWEL 1992, 69–74), FIGURE 5.8

Eastern Greek chariots vary slightly from their mainland companions. The evidence for eastern Greek chariots is even more limited than that for mainland Greece, but several differences can be noted. Representations of chariots are limited to a period from the second half of the sixth to the early fifth century BCE, and appear only in painting and terracotta relief plaques. The axle seems to be fixed as in the mainland examples, although the axle position is placed farther to the rear of the body. Decorative axle caps, which hide the linchpins, are borrowed from other eastern neighbors. The most distinctive difference is that eastern Greek chariot depictions more commonly feature six- or eight-spoked wheels than the typical four-spoked version in mainland Greece (fig. 5.8).[35]

Eastern Greek chariots bore a greater resemblance to the contemporary Near Eastern, Achaemenid, and Assyrian chariots than did the mainland versions. The geographical situation of Ionia with ties to both the East and West makes this area a veritable melting pot for ideas, technology, and artistic expression. Figure 5.8 is a mold made of architectural terracotta, which would have covered the wooden architrave of a post-and-beam building. The details of the plaque would have been brightly painted and legible from the ground.

The fragmentary plaque preserves a profile view of a chariot moving left, pulled by two horses. Two figures stand in the chariot body: a charioteer holding the reins, and a warrior dressed in armor and outfitted with a spear and

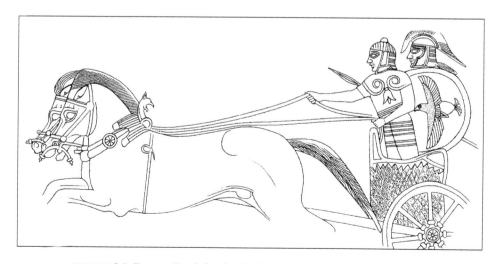

FIGURE 5.8. Eastern Greek fixed-axle chariot circa 550 BCE, after Crouwel.

shield. The body of the chariot features an elaborately decorated surface. The wheel has eight spokes, a covered axle end, and a composite felloe. The spokes are single-piece, not bipartite as seen in mainland representations. The spokes are, however, elaborated with a knob near the nave. It is not clear whether this adds strength or is merely decorative, but parallels are present in Achaemenid wheels. The axle end is decorated with a rosette cover that obscures any indication of the linchpin. The coroplast has rendered the felloe with two concentric circles, and presumably the exterior is a tire of either rawhide or wood.

The spokes commonly show decorative carving or turning, another trait borrowed from Achaemenid chariots. There is insufficient evidence to determine tire presence or type. A three-quarter view from an Etruscan vase painting gives a clearer view of the relationship of multi-spoked wheel nave to axle in figure 5.9.[36]

Although late classical Etruria is out of our sphere of interest, this vase painting depiction of a chariot provides evidence for the relationship of the

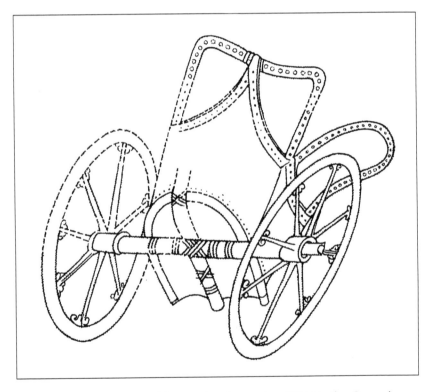

FIGURE 5.9. Etruscan red-figure volute krater circa 375 BCE, after Crouwel.

wheel to the axle true of most Greek chariots. The three-quarter, underside view shows the bottom of the chariot body. A draught pole extends vertically from the back of the chariot body forward to the team of horses. Traction for the chariot is provided by the horses attached to the draught pole by means of yokes.[37] The axle crosses under the draught pole horizontally, and the two appear to be bound together and to the chariot body by means of thongs.

The depiction of the far wheel shows the cylindrical nave slipped over the axle as a collar. The nave appears to be slightly larger in diameter than the axle and features a ring element, possibly a bushing or constricting band, on the interior face. The depiction of the near wheel clearly shows the axle projecting out from the nave, and there may be a suggestion of a linchpin. In the painting, the projecting axle end appears to be smaller in diameter than the main portion of the axle under the chariot body, as was observed in figure 5.6.

The spokes on this Etruscan example attach at the felloe with decorative wedges. The spokes appear to be single, not bipartite, although not turned decoratively as seen in the eastern Greek example (fig. 5.8). The spokes appear to narrow from nave to felloe where small volute-like wedges strengthen the join. These small supports are in contrast to the large triangular or scalloped versions found on mainland chariot wheels (see fig. 5.5). The spokes are mortised into the cylindrical nave as seen in figure 5.7.

C.2.C

FIXED AXLE: CARTS AND WAGONS, FIGURES 5.10 AND 5.11

Again, there is little surviving archaeological evidence for carts and wagons, but the visual evidence includes numerous depictions in painting, on coins, in relief, and in three-dimensional figurines.[38] Some faster-moving transport vehicles also had fixed axles. In depictions of carts, we can recognize the fixed axle with its round end (cf. rotating axle below). In addition to speed, the fixed axle permitted differential turning of the wheels to facilitate maneuvering over rough terrain. Most are shown with crossbar wheels, but there is evidence for spoked and disk wheels. The crossbar wheel is simpler but not as strong as the spoked wheel. It was also probably adopted from the East. A crossbar wheel of oak and beech wood is known from Olympia, dated to the first quarter fifth century BCE, shown in figure 5.10.[39] Again, had Anaximander imagined the map of the earth in the tradition later articulated by Aristotle's *Meteorology*, the visual presentation evokes a sense of a tripartite map, this time incorporating the fixed axle.

Slower-moving vehicles, carts or wagons, could have fixed or rotating axles. This example of a crossbar wheel has a circular nave to accommodate a fixed axle. As with the chariot, the fixed axle ran beneath the body of the

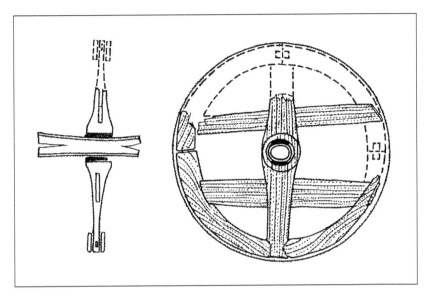

FIGURE 5.10. Fragmentary crossbar wheel from Olympia, early fifth century BCE, after Crouwel.

cart, attached to the underside. This crossbar wheel from Olympia, made of oak, has a reconstructed diameter of 0.78 m. The cylindrical nave has a reconstructed length of 0.375 m, with an inner diameter of 0.06 m. The interior of the nave shows signs of wear, thus confirming a fixed axle. A horizontal iron band reinforced the wooden nave against the pressure of the axle. These bands are occasionally seen in the visual representations of crossbar wheels. The wheel would have been held on with a linchpin.

The crossbar wheel has a single diametric bar through which the axle passes and two or more perpendicular crossbars join to it. The Olympia example features a diametrical crossbar made of two separate pieces joined with the nave, all made of beech wood. The pieces widen toward the nave, have mortises to accept the tenons of the other crossbars, and join with the felloe with mortises. The felloe (reconstructed height 0.09 m) was made of four segments of bent wood mortised together. There is no evidence for a tire.

The fixed axle for slow-moving carts permitted more maneuverability than a rotating axle. The independently moving wheels adjusted differentially to the terrain, and allowed for easy turning. The crossbar wheel has the advantage of being quite simple and sturdy under heavy weight. It is less sophisticated than the spoked wheel and easier to repair by unskilled workers. Crouwel suggests that the origin of the crossbar wheel is likely the Near East, as it is known there earlier than in Greece.[40]

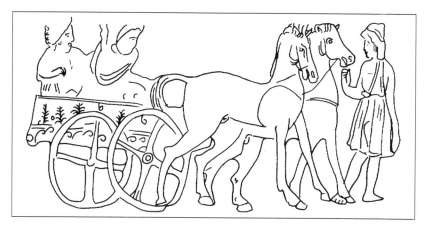

FIGURE 5.11. Fixed axle with spoked wheel. Attic red-figure pyxis, from Eretria circa 400 BCE, after Crouwel.

The nave of this wheel had an iron hoop at one end to prevent it from splitting. Spoked wheels from a cart are known from a sixth-century BCE burial near Larissa in Thessaly. These wheels bore iron tires, and the cart had other iron fittings. Metal reinforcements must have been more important for weight-bearing vehicles than for the lightweight racing chariots. Figure 5.11 shows a cart in use.[41]

Carts with fixed axle and spoked wheel clearly follow the structural design of the chariot. The axle passes under the cart body; the naves of the spoked wheels fit over the axle; and the wheel is secured with a linchpin through the projecting axle end. The difference is in the body of the vehicle itself. Advantages of this system are speed, maneuverability, and lightness. Carts with spoked wheels may not have been as sturdy and able to bear as large cargoes as crossbar wheeled carts. Wheels are always shown with four spokes.

In addition to this late classical red-figure depiction of a spoked cart wheel, there are surviving wheels from a sixth-century BCE burial near Larissa in Thessaly.[42] The wooden wheels do not survive, but their iron fittings do. Their reconstructed diameters are 0.80 and 1.10 m. Smaller iron rings (d. ca. 0.10 m) may be nave hoops as seen on the Olympia crossbar wheel example (fig. 5.10). These wheels may have belonged to hearses.

C.2.D

ROTATING AXLE: CARTS AND WAGONS
(CROUWEL 1992, 77–87), FIGURES 5.12–5.13

The rotating axle is seen in depictions of wheels without round axle ends. The fixed axle is round in the section under the floor of the cart, but rectangular where the wheels attach. The axle joins the cart bottom with a series of brackets of an inverted U shape, as seen in figure 5.12.[43]

In contradistinction to earlier displays that might have captured the inspiration in Anaximander's tripartite map, had Anaximander imagined the tripartite divisions in terms of the measurement of the short shadow marker on the summer and winter solstices, the resulting map would have more clearly resembled column-drum *anathyrôsis*, the imagery of which is suggested by these wheels on the cart model. We shall explore this possibility in chapter six.

The alternative to a fixed axle is a rotating axle. In this scenario, the wheels are attached to the axle, and both wheels and axle rotate as one. On this fragmentary depiction of a cart, a cross-bar wheel is attached to a rotating axle. The axle end appears as a rectangular tenon passing through the diametrical crossbar of the wheel. The wheel was secured by a linchpin. The axle and wheels are a single unit, and they are not permanently attached to the cart. In this depiction, the axle rides in an axle bracket of inverted U form. The cart could be lifted off the axle/wheel unit. This system would suit slow-moving vehicles best, and provides easy disassembly for repair or storage.

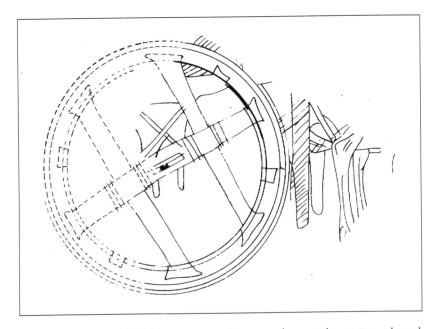

FIGURE 5.12. Attic black-figure amphora fragment of cart with rotating axle and cross-bar wheel, after Crouwel.

The crossbar wheel in this black-figure example (fig. 5.12) resembles the Olympia crossbar wheel (fig. 5.10). In this case the diametrical cross-bar does not form the nave, but features the rectangular slot for the axle end at its center. The perpendicular crossbars are further secured with bindings along the diametrical crossbar. The crossbars are mortised into the felloes, although it is not possible to tell if the felloe is made of multiple segments. The felloe is rendered with several concentric circles suggesting reinforcing iron rings or tires. Several wedges are shown along the interior of the felloe, and it is possible that these are iron felloe clamps holding a tire to the felloe.[44]

Thus, the cart body is not physically attached to the axle and can be lifted off the axle with wheels if necessary. Both wheels and the axle will rotate simultaneously, thus providing little maneuverability but much stability. The wheels are most often cross-bar, but spoked and disk wheels are also known. Archaeological remains of rotating axle crossbar wheels are known from Gordion (sixth century BCE) among other sites.[45] There is no archaeological evidence for tires, but wheels are shown with a composite felloe suggesting either a rawhide tire, another wooden rim, or a sweated on iron tire. Disk wheels are not attested archaeologically, but they appear in representations of carts, as in figure 5.13.[46] It is difficult to know how literally to take these representations and what their relationship to actual, functional carts is.

The evidence for disk wheels in mainland and eastern Greece is limited. Disk wheels rarely appear in vase painting, and when they do they are featured on children's wagons or toys.[47] They appear more often on three-dimensional cart and wagon models. The difficulty is assessing the accuracy of the stylized and coarsely formed models without supporting evidence from archaeology or other media.

The cart model illustrated here (fig. 5.13) features four disk wheels painted with concentric circles on both sides. The body of the cart imitates the body of a horse, and a horse head appears at the front of the cart. In the model, the axle rotates with the wheels. The relationship of this model and actual contemporary carts is difficult to assess. The horse form seems unlikely, and the model may be alluding to and abbreviating the horse required to pull the cart by integrating it into the cart body. On the cart are six banded amphorae. This cart was a grave offering, and probably makes a reference to the bounty and prosperity of the dead person's life. As an object of the late geometric period, the decorative motives are linear and monochromatic. The concentric rings may be purely decorative, but one can guess that actual carts received decoration in the same linear, geometric tradition. The small scale and media of the models may explain the presence of disk wheels. Disk wheels are much easier to make than spoked versions in terracotta and bronze; thus, it is possible that the disk form of the wheel is only a product of

FIGURE 5.13. Attic black-figure amphora fragment of cart with rotating axle and crossbar wheel, after Crouwel.

the medium, not an intentional representation of an actual cart detail. This is particularly true of a bronze cart model from Çesme (near Smyrna), which lacks many distinguishing details in favor of a generalized form.[48]

C.3

ANAXIMANDER AND THE CHARIOT WHEEL REVISITED:

Cosmic Wheel and Axle, Cosmic Tree, and Axis Mundi

Because Anaximander imagined the sun, moon, and stars as circular wheels and also imagined the shape of the earth, in plan view, as a circle, an examination of wheel making and its attendant technologies supplies much for further consideration. What we have then are three issues that bear directly on our reflections of Anaximander's cosmic wheels in light of our exploration of practical wheel making: (1) Did Anaximander witness any wheels with hollow rims that might have supplied an inspiration for his imagination? (2) Did

Anaximander's attention to wheel making allow him to focus on wheels connected inextricably to an axle and so imagine the earth, stars, moon, and sun connected by a cosmic axle, that is, an *axis mundi*? (3) Did wheel making, whether rectangular techniques or the concentric circles displayed by models, supply some clues by means of which Anaximander imagined his tripartite map? We will explore further the issue of the hollow-rimmed wheels and the cosmic axle; the issues surrounding the map will be explored later when we turn to consider his seasonal sundial.

Our survey of chariot wheels and their operation is derived mainly from the visual representations of vehicles. To acknowledge the source of evidence is at once to acknowledge what tends to be misleading about it. The iconography of late archaic and early classical vase painting favored the heroic both mythologically and in the elevation of mortal life to a higher plane through assimilation to the world of heroes. Chariots with their elaborate and sophisticated spoke wheels are an element of the heroic, invoking military deeds of the warriors of the past. The use of chariots in contemporary culture was possibly limited to ceremony and ritual, including races and athletic contests, and had no prominent place in contemporary military tactics. But the best argument for the identification of cosmic wheels with a chariot follows from the fact that chariots routinely made use of fixed axles, and Anaximander's motionless earth is captured by the spirit of the fixed-axle construction. On the other hand, wheeled carts and conveyances must have populated the streets and roads of the city and countryside, and some of these also had fixed axles. These humble vehicles hold only occasional interest to the artists of the archaic and classical period, thus limiting their presence in the visual evidence. We must recall this critical role of the artist as filter for life when we analyze the images from any period.

In our survey of wheel forms, we found a variety of wheel constructions available to the archaic wheelwright, and to Anaximander's inspection. The spoke wheel on a chariot is the most commonly represented. Less frequently represented, but present in images and models of wheeled vehicles, are less sophisticated wheel constructions associated with carts and slow-moving vehicles. These slow-moving vehicles frequently featured rotating axles—spoke, planks, and disk. What these carts lacked in speed and maneuverability, they made up for in durability and maintainability. Construction of the chariot wheel required skilled workmanship to manipulate the wood, design and install the spokes, and ensure balance. In some cases, metalworking was also required to fit out tires or reinforcing bands. The conjunction of metalworking and woodworking may have occurred in the same establishment, such as a cooper's workshop, but the archaeological evidence for urban "industrial districts" suggests that close proximity of various crafts could result in cooperative exchange of labor. In the case of the rotating axle wheels, the construction appears to be more vernacular and less skilled. These are wheels that may be wrought by a local, village craftsman or even the cart owner him-self. The wear on carts from rough roads and heavy loads certainly resulted in frequent repairs to worn or damaged wheels, so the simplicity of construction and design favored the moderately skilled person since the repairs could be expected to be carried out by such an individual.

For the observer of a wheel in motion, the key elements are the rim (fel-loe) and nave/axle. The spinning of a spoke wheel adds the mesmerizing and dizzying blur of moving forms. Pindar uses this visual unsteadiness as a metaphor for drunkenness, thus emphasizing the ability of motion to trans-port one to another level of consciousness. On the other hand, the static and central axle and nave would convey a sense of order amid the blur. Indeed, the very functioning of the wheel depends on the stability of the hub amid the movement toward the periphery. Since Anaximander, according to these late tertiary reports, called on the imagery of the chariot wheel to illuminate the heavenly wheels, we are entitled to supply this background information to reveal the context of his proposal. The background includes the wheel-wright's technology, along with the proliferation of the variety of wheeled vehicles he and his compatriots would have seen, and the perception of a spinning wheel itself. The reference to the chariot wheel is both simple and isolated, and, like the reference to the column-drum earth, it points believ-ably to a complex scene of technology and industry that informed Anaxi-mander's cosmic imagination. There is no evidence, for example, that Anax-imander held that the earth is spinning, and so the imagery of the spinning cosmic wheels around the central earth directly lends itself to the analogy of the earth with the nave and axle. The axle could be seen as a metaphor for the *axis mundi* supporting the spinning wheels in the heavens, and we shall explore this tempting possibility directly. While the surviving testimony only mentions the wheel metaphor, we know that Anaximander wrote a book, and it is tempting to suppose that his book supplied more technological details. There can be no doubt that Anaximander found this imagery fitting to illu-minate the cosmic structure and its mechanism. Let us return to our opening investigation and reassemble the pieces of the puzzle.

At the beginning of this chapter, we set out a range of problems that challenged our understanding of Anaximander's cosmic wheels. Why did Anaximander find it suitable to explain the sun, for example, by appeal to the image of a wheel? Certainly, an immediate and important reply is that the wheel image helped to explain why the "sun" did not fall, as would be expected if it were instead a heavy object. Did the appeal to the wheel imagery not only clarify the structure of the appearance but also supply a hint at an explanation of how the cosmic wheels moved? If the mechanism of the motion of the wheels was part of Anaximander's imagination, was a cosmic axle presupposed to ensure the uniformity of heavenly movement? Moreover, do any of the archaic examples supply the essential imagery of a *hollow* felloe that could be filled with fire—*hamarteiô trochô paraplêsion, tên hapsida echonta koilên, plêre puros?*[49] Finally, did the wheelwright's shop possibly provide a way for Anaximander to think about his seasonal sundial, a device that seeks to reveal the mechanical patterns made by the sun's spinning wheel?[50]

From the wheelwright's shop, there is one remarkable example of a special wheel-and-axle combination that appeared at the temple building site connected indispensably with the movement of architraves and even ceremonial columns.[51] In this special wheel-and-axle combination, perhaps we can find an answer to the perplexing image of a (chariot) wheel with its felloe *hollow.*[52] Unless the wheel is hollow, it cannot be full of fire. While an inspection of the wheels that we have so far described turns up no likely candidate, the wheelwright in conjunction with the architects produced such an example. We know that the ancient architects were described as technical polymaths who planned, built, sculpted, casted bronze, invented tools, and worked with miniatures. One famous example mentioned by Pliny credits the architect Theodorus with making a chariot pulled by four horses that was so tiny it could be hidden behind the wing of a fly. In another case, Vitruvius tells us that an architect produced vehicles of transport with a fixed axle; we know on good authority that Chersiphron and Metagenes, the architects of Artemision C, the marble temple to Artemis in Ephesus, probably begun not long after 560 BCE, were credited with inventing special vehicles of transport and publishing a book about it. The creation of new wheeled vehicles was necessary not only because of the great distance from the quarries to the building site but, more importantly, due to the increased weight of the marble monoliths. Chersiphron's invention was to create a great wheel, with the large block or column drum lodged at the center as its nave, to be pulled by oxen. His son Metagenes was credited with extending this idea to deliver the lengthy architraves as well. In Metagenes' technique, the architrave block, and/or ceremonial monolithic columns, became the axle between two large Chersiphron-styled wheels. Vitruvius describes in detail Chersiphron's invention, a special vehicle for transporting column drums:

> It may not be out of place to explain the ingenious procedure of Chersiphron. Desiring to convey the shafts [sc. column drums] from the temple of Diana of Ephesos, from the stone quarries, and not trusting to carts, lest their wheels should be engulfed on account of the great weights of the load and the softness of the roads in the plain, he tried the following plan. Using four-inch timbers, he joined two of them, each as long as the shaft, with two cross-pieces set between them, dovetailing all together, and then leaded iron gudgeons shaped like dovetails into the ends of the shafts, as dowels are leaded, and in the woodwork he fixed rings to contain the pivots, and fastened wooden cheeks to the ends. The pivots, being enclosed in the rings,

turned freely. So when yokes of oxen began to draw the four-inch frame, they made the shaft revolve constantly, turning it by means of the pivots and the rings.[53]

Immediately following, Vitruvius turns to explain how Cherisphron's son, Metagenes, developed this technique to deliver the heavier and larger architraves to the building site:

> When they had thus transported all the shafts, and it became necessary to transport the architraves, Chersiphron's son Metagenes extended the same principle from the transportation of the shafts (sc. column drums) to the bringing down of the architraves. He made wheels, each about twelve feet in diameter, and enclosed the ends of the architraves in the wheels. In the ends he fixed pivots and rings in the same way. So when the four-inch frames were drawn by oxen, the wheels turned on the pivots enclosed in the rings, and the architraves, which were enclosed like axles in the wheels, soon reached the building, in the same way as the shafts.[54]

Consider the reconstruction of Metagenes' invention shown in figure 5.14, after Coulton.[55] What we have here is a wheel with its felloe hollow; the hollow empty space in each wheel is created as a result of the extended width of the wheel. Vitruvius mentions Metagenes' invention along with crediting

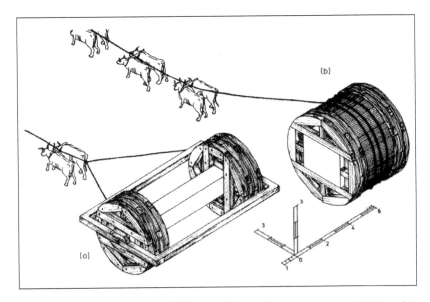

FIGURE 5.14. Reconstruction of Metagenes' invention to move monolithic architraves (a), and a similar technique to move large blocks and drums (b), after Coulton.

him and Chersiphron with writing a prose book. It seems plausible that such a book would have contained a sketch of this device; perhaps Vitruvius saw it himself when detailing it in his prose book. And had there been such a sketch, among others, it would make more likely still that Anaximander's contemporaneous book contained informal sketches, perhaps with numbers attached. Thus, it may very well have been that a wheel after this fashion inspired Anaximander to imagine his heavenly wheels. It too made use of a fixed axle. Although this wheel is certainly not a chariot wheel, the late report might very well be garbled; then, Anaximander had imagined the cosmic wheels, each one earth diameter in width, full of fire, connected by some kind of axle. With this imagery in mind, let us reflect further on Anaximander's cosmos. Anaximander imagined the formation of the cosmos with an original fire surrounding it *like bark around a tree*. When we do so, the idea of a cosmic axle appears in connection with a cosmic *tree*.

Let us recall that in explaining the origins and development of the cosmos from its beginning, Anaximander held that a great fire surrounded the cosmic extremities while a cold and wet earth stood at the center. That original fire is sometimes identified as a sphere of flame, but this too seems like a misreporting because the fire surrounded like *bark around a tree*.[56] Anaximander's cosmos is *cylindrical* like a column, not spherical; the cosmos is alive—hylozoistic—and the organism grows like a great tree. Imagined in plan view, or what proves to be a horizontal cross section through the plane of the earth, the cosmos displays concentric circles like rings in a tree. Recall also that in the temples of the seventh century BCE the roofs were held up by trees and were replaced only when the temples were widened and the available timber could no longer carry the increased load of the enlarged roof. The stone column replaced the tree; the column drum was a piece of the original tree and its symbolic meaning was attached inextricably to it, now as part of a petrified forest. Thus, if we think through the imagery, the column-drum earth is situated in the middle, a piece of the original tree—in temple and in cosmos—and from that we can imagine a tree-trunk axis, as invisible as the heavenly wheels themselves, extending above, below, and through the column-drum earth. It is hard to imagine how a Greek of the sixth century BCE could have imagined a cosmic *wheel* without also imagining an *axle* connecting to it.

The tree imagery and its cosmic role have a long documented history.[57] In Hesiod, we are told of the roots of the earth that extend downwards into a seemingly bottomless chasm.[58] A similar sentiment and imagery are found in the Orphic Rhapsodies,[59] and Xenophanes too refers to the earth rooted in the seemingly bottomless depth below.[60] The rooted earth is also known to Aeschylos,[61] Kallimachos,[62] and Apollonios of Rhodes.[63] The conception of a "rooted earth" emerges only when the earth itself is imagined as a large tree; this very conception is embraced by Pherecydes, who developed this imagery further by imagining a "winged earth" (hypopteras).[64] Thus, from the time of Homer onwards, the culture is drenched in descriptions of tall trees, pines that are heaven high, and lofty-headed oaks.[65] The Greek tradition itself, then, apart from cosmic trees imagined by Iranian and Babylonian neighbors, imagined "a cosmic tree whose roots extend into the underworld, whose trunk forms the main body of the earth and sea, and whose branches reach into the heavens."[66] West[67] and Schibili conjectured that the tree may be thought of as supporting and providing a frame for different parts of the cosmos.[68] Anaximander's mentality needs to be reviewed within just this cultural context; his conception presupposes a world axis constituted by the trunk of this great tree and, while not calling on Pherecydes' winged tree, his "tree-structured" cosmos has rotating cosmic wheels. What Anaximander added to this tradition was a consequence of imagining cosmic movement as analogous to rotating wheels, probably with a fixed axle. Anaximander imagined the cosmos as alive, growing, in cylindrical shape like a great tree; its branches are cosmic wheels united by an *axis mundi*, the physical fact of which is the column-drum earth. An *axis mundi* is the central pivot of the earth or the entire cosmos.

In Plato's cosmology, the spherical earth was surrounded by a crystalline sphere of fixed stars, supported by an *axis mundi*, that Plato has Er call "the Spindle of Necessity" (anankês atrakton).[69] This Spindle of Necessity is a shaft of light running straight through heaven and earth like a *column* (kiona). Whatever else it might be, it is the *axis mundi*, and takes the shape of a column; this conception traces a lineage of traditions that should include Anaximander, for it is Anaximander who explicitly interweaves the tradition of the cosmic tree and cosmic column.

The long historical picture of traditions holding that there is, in fact, a cosmic tree connecting earth to heaven was summarized succinctly by Schibili:

> The cosmic tree, be it in Norse mythology, Babylonian creation accounts or in the cosmology of Pherekydes of Syros, seems to have had a long and varied history. From earliest times primitive people in Europe and Asia erected pillars or towers (Weltsäulen) that were to stand for an imagined column at the center of the world; this central column supported the dome of heaven and around it the firmament revolved. Often these pillars would take the shape of trees.[70]

Once we allow our reflections to bring together cosmic tree and column, and return to consider Metagenes' vehicle of transport whose wheels do provide some sort of hollow rims, we can find yet another powerful and suggestive image when the wheel-and-axle setup is placed alongside an elevation view of Anaximander's cosmos (fig. 5.15). Placing these two images side-by-side, we can see Anaximander's cylindrical cosmos—each wheel adding to the rings in the cosmic tree—forming an *axis mundi*.

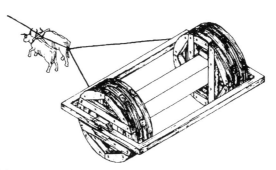

(a) Metagenes' oxen-drawn wheeled vehicle for transporting monolithic architraves.

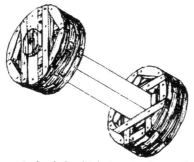

(b) Metagenes' wheeled vehicle imagined as in elevation.

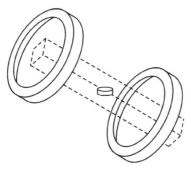

(c) Anaximander's cosmos in elevation; the circles representing the wheel of the sun at summer and winter solstices.

FIGURE 5.15. Anaximander's imaginative projection of the architect's wheeled vehicle to represent the sun wheel at summer and winter solstice, suggesting a cosmic axle, an *axis mundi*.

SIX

Anaximander's Cosmic Picture

Reconstructing the Seasonal Sundial
for the Archaeologist's Investigations

A

THE DOXOGRAPHICAL REPORTS

Ἀναξίμανδρος Πραξιάδου Μιλήσιος . . . εὗρεν δὲ καὶ γνώμονα πρῶτος καὶ ἔστησεν ἐπὶ τῶν σκιοθήρων ἐν Λακεδαίμονι, καθά φησι Φαβωρῖνος ἐν Παντοδαπῇ ἱστορίᾳ, τροπάς τε καὶ ἰσημερίας σημαίνοντα, καὶ ὡροσκοπεῖα κατεσκεύασε. καὶ γῆς καὶ θαλάσσης περίμετρον πρῶτος ἔγραψεν, ἀλλὰ καὶ σφαῖραν κατεσκεύασε.

—Diogenes Laertius II, 1–2; DK 12A1]

Anaximander, the son of Praxiades, of Miletos, was first to discover the gnomon and set it up on the sundials in Sparta, according to what Favorinus says in Universal History, in order to mark the solstices and equinoxes, and also he constructed markings to show the hours. He was the first to draw an outline of the earth and sea, and he also constructed a (celestial) model.

Ἀναξίμανδρος Πραξιάδου Μιλήσιος . . . πρῶτος δὲ ἰσημερίαν εὗρε καὶ τροπὰς καὶ ὡρολογεῖα, καὶ τὴν γῆν ἐν μεσαιτάτῳ κεῖσθαι. γνώμονά τε εἰσήγαγε καὶ ὅλως γεωμετρίας ὑποτύπωσιν ἔδειξεν. ἔγραψε Περὶ φύσεως, Γῆς περίοδον καὶ Περὶ τῶν ἀπλανῶν καὶ Σφαῖραν καὶ ἄλλα τινά.

—*Suda* s.v.

Anaximander son of Praxiades, of Miletos, first discovered the equinox and solstices and hour-markers, and that the earth lies in the center. He introduced the gnomon and made known an outline of geometry. He wrote "On Nature," "Circuit of the Earth," and "On the Fixed Stars," a "Celestial Globe," and some other works.

Ἀναξίμανδρος ὁ Μιλήσιος ἀκουστὴς Θάλεω πρῶτος ἐτόλμησε τὴν οἰκουμένην ἐν πίνακι γράψαι. μεθ᾽ ὃν Ἑκαταῖος ὁ Μιλήσιος ἀνὴρ πολυπλανὴς διηκρίβωσεν, ὥστε θαυμασθῆναι τὸ πρᾶγμα.

—Agathemerus, I.1

Anaximander the Milesian, a disciple of Thales, first dared to draw a map of the inhabited earth on a tablet; after him Hecateus the Milesian, a much-traveled man, made the map accurate enough to become a thing of wonder.

. . . τοὺς πρώτους μεθ᾽ Ὅμηρον δύο φησὶν Ἐρατοσθένης, Ἀναξίμανδρόν τε Θαλοῦ γεγονότα γνώριμον καὶ πολίτην καὶ Ἑκαταῖον τὸν Μιλήσιον. τὸν μὲν οὖν ἐκδοῦναι πρῶτον γεωγραφικὸν πίνακα, τὸν δὲ Ἑκαταῖον καταλιπεῖν γράμμα πιστούμενον ἐκείνου εἶναι ἐκ τῆς ἄλλης αὐτοῦ γραφῆς.

—Strabo I, 7

Eratosthenes reports that the first to follow Homer were two, Anaximander who was an acquaintance and fellow-citizen of Thales, and Hecateus the Milesian. Anaximander was the first to publish a geographical map, while Hecateus left behind a drawing believed to be his from the rest of his writing.

B

THE SCHOLARLY DEBATES OVER
THE TEXT AND ITS INTERPRETATION

WE HAVE TWO reports that Anaximander made a *seasonal* sundial; a seasonal sundial would be a shadow-catching instrument that identified the first day of summer and winter (i.e., the solstices) and the first day of spring and fall (i.e., the equinoxes). One report comes from Diogenes Laertius;[1] on the authority of Favorinus, Anaximander discovered the gnomon and set one up on the sundials in Sparta to measure the solstices and equinoxes, and it also had hour indicators. Another report comes from the *Suda*;[2] Anaximander discovered the equinox, solstices, and hour indicators, and he introduced the gnomon. The discovery or introduction of the gnomon is reported in the same breath as the discovery and measurement of the sun's turnings (solstices) and the point when there is equal day and equal night (equinoxes). And it is of great importance to note also that in the very same passage Anaximander is credited with making a map of the world and also a model or drawing of the cosmos. Diogenes Laertius credits him with making a *perimetron* of the earth and sea and also a *sphairos*, while the Suda mentions *gês periodon* (a work on the map of the earth) and a *sphairos*. And, in two other passages, Agathemerus credits Anaximander with making a *pinax* of the *oikumenê*,[3] and Strabo mentions also the *pinax* that was perfected further by Hecataeus.[4] So, what we have are reports that Anaximander invented a seasonal sundial connected to some introduction or invention of a gnomon and, in the same passages, supplanted by other reports, that Anaximander made a map of the earth and some kind of model or drawing of the cosmos. The most strident rejection of the doxographical reports that credit Anaximander with inventing a seasonal sundial, and discovering or measuring the solstices and equinoxes by means of it, comes from Dicks, who dismisses it all wholesale. He refuses to accept the accounts because he doubts that Anaximander could have understood the concept of the equinoxes, the use of a gnomon to measure them, and familiarity with the concept of the celestial sphere that he regards as requisite to making such measurements.[5] This wholesale dismissal is rejected by most commentators, though some of the doubts are shared. Can a plausible case be constructed that Anaximander indeed made a seasonal sundial? How might he have made it? And when we "reconstruct" it, what will we discover about his understanding of astronomical phenomena and the applied geometrical techniques that were revealed by his making of such a sundial?

The hypothesis I propose to explore is that sundial, gnomon, earthly map, and cosmic model were all integrally connected; indeed, in Sparta all these ingredients plausibly came together to make one and the same thing— a seasonal sundial, map, and model with a specially measured gnomon set in the center. Let us explore these ideas.

C

RECONSTRUCTING THE SUNDIAL
FOR THE ARCHAEOLOGIST'S EXPLORATIONS

Unlike previous chapters, where a review of the scholarly debates is followed by the presentation of the archaeological evidence, we have no claims, thus far, that archaeologists have found the seasonal sundial referred to in these doxographical reports. We do have archaeological evidence for ancient sundials, earlier ones outside of Greece and later ones inside Greece dating to the Hellenistic period, and these resources will come to play a vital role in this chapter. But the orientation here is to propose a reconstruction of Anaximander's seasonal sundial, that is, to propose the details of how it looked, so that archaeologists might search for this sundial, or perhaps even recognize that they have already found it but that it has somehow escaped their notice.

In *Anaximander and the Architects*,[6] I attempted to reconstruct Anaximander's sundial following the suggestion by Sharon Gibbs in *Greek and Roman Sundials*.[7] Gibbs suggested that the markings found inside the north wall of a well on Chios, although too late to be Anaximander's design, displayed the essential ingredients of a "seasonal" sundial. With a gnomon, pointing south, set *horizontally* into the north wall of the well, the deepest shadow in the well marked the summer solstice and the shortest shadow marked the winter solstice. Having reconstructed the markings in the well, it was clear also that the "equinox" shadow was not midway between the two but rather closer to the winter solstice marker, as shown in figure 6.1.

I am in agreement with Gibbs that these are relevant markers for the sundial attributed to Anaximander. But, after working for a number of seasons with students in Greece making sundials, it became clear to me that the design was mistaken. While it is true that, as Gibbs shows, the earliest surviving Greek sundials date to Hellenistic times, and their designs are either hemispherical or conical, it seems more fitting for Anaximander to have used a gnomon set *vertically*, not horizontally; in the process of reflecting upon a design appropriate both to Anaximander and to the sixth century BCE, I now propose a new hypothesis of the design Anaximander made.

In light of a review of the reports mentioning together the gnomon, the sundial, the map of the earth, and the model of the cosmos, let us begin again by thinking of Anaximander's sundial face as a column drum. Why? Because Anaximander had described the earth as a flat disk, analogous with it,[8] it might well have dawned on him that the shadows cast on the column drum would be analogous to the shadows cast on the earth itself, a microcosm of the macrocosm.[9] And furthermore, as will become clearer, the sun-shadow markings would provide clues for the map with which Anaximander is also credited, since some sundial markings were key to the making of a "frame" for it. And since the shadows marked on this *skiatherion* were caused by the sun, Anaximander's cosmic picture could be inferred by means of those shadow markings. *The sundial, then, holds the key to connecting the earthly map and the cosmic map.*

Now, let us explore this construction further by imagining that in the center of a column drum, where the architect's *empolion* would be, Anaximander placed a gnomon, set vertically.[10] Let's now imagine plotting the *shortest shadow each day*, which we can call "local noon." With a gnomon set vertically, the "shortest shadows" marked each day form a straight line running north–south. If we make a line perpendicular to those shadow markings, we not only create a diameter by bisecting the (column-drum face) circle but the line will mark due east–west and consequently will establish the sighting

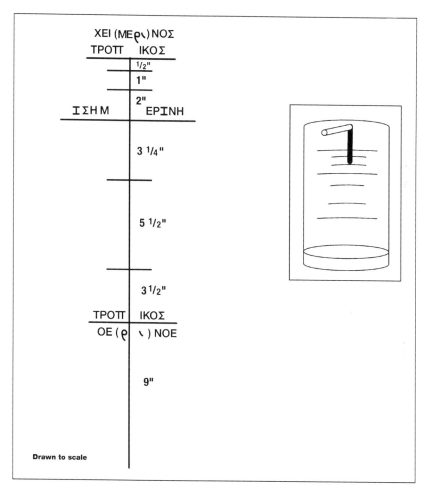

FIGURE 6.1. A possible reconstruction of Anaximander's sundial with the gnomon set horizontally pointing southward and casting shadows on the north wall of a well in Chios.

points for the rising and setting of the sun on the equinoxes. Thus, for Anaximander, the local noon shadow markings will appear in the "northern half of the semicircle" of that circular column-drum face, perpendicular to a straight-line diameter bisecting the column-drum face running east–west (fig. 6.2). To realize, and moreover to *confirm*, the geometry of this astronomy requires careful observations, over the course of time, whether or not the original construction was done as a geometrical figure.[11] Only by careful observation could the success of the device be confirmed.

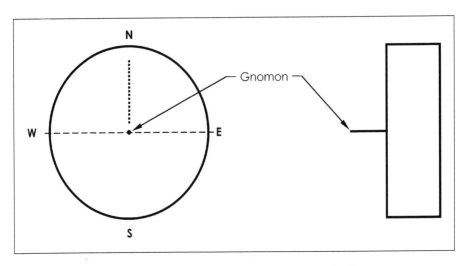

FIGURE 6.2. Circle/column-drum face bisected east–west with many local noon points running north–south.

The *shortest* of all the short shadows over the course of the whole year marks the summer solstice, and the *longest* of the short shadows marks the winter solstice. When we identify the equinox by modern methods, it is plainly clear that the shortest shadow on the day of the equinox is *not* midway between the shortest and longest "local noon" markers—it is closer to the shortest of the shadows (i.e., it is closer to the summer solstice marker, and not closer to the winter solstice marker as it is when the gnomon is set horizontally—see fig. 6.3). So, it remains for us to consider how Anaximander might have reckoned exactly *when* the equinox occurred, if indeed he had done so.

Now, let us turn for a moment to begin a consideration of Anaximander's map. This is complicated slightly, first of all, because Agathemerus and Strabo report that he made a *pinax* of the *oikoumenê*, while Diogenes says that he made a *perimetron* of the earth and seas.[12] The complication is that, on the one hand, he made a map of the inhabited earth and, on the other, he made a map of the earth that would certainly include uninhabitable regions, and so quite a different map.[13] But, for our purposes here, let us just consider some outlines on the column-drum face to supply the reference frame for the map. A conventional approach has taken its lead from Heidel, namely, that to make a map one needs to make a "frame," and this is done by constructing something of a rectangle using the rising and setting points of the sun, against the horizon, on the summer and winter solstices, and the equinox.[14] Heidel took up Aristotle's explanation in the *Meteorologica*[15] of

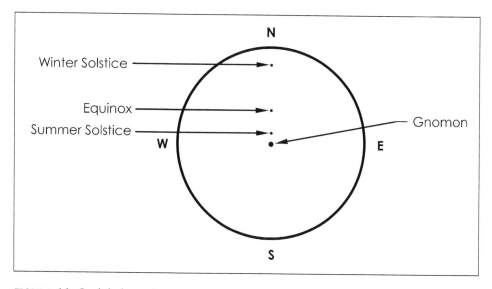

FIGURE 6.3. Circle/column-drum face with only three shadow points: summer solstice, equinox, and winter solstice.

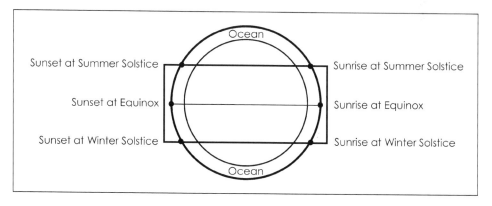

FIGURE 6.4. Anaximander's Earth with sun risings and settings on solstices and equinox, and rectangle suggesting the inhabited earth.

just these markers to make an earthly frame, and, in a simplified form, this idea is illustrated in figure 6.4.

The key element that has been neglected in the scholarly discussion is the reckoning of the *center* as well as the outside frame. Once considered, it is obvious that the whole picture is transformed. And so the question remains: How did Anaximander conclude where the center was on this flat-disk earth?

How did he determine where *he*, the observer, was on the flat disk? And how did he reach those conclusions? The answer proposed here is that his sundial revealed it to him, and in turn he revealed it to his community in a book that almost certainly discussed it.

Do we have any surviving examples that resemble the sundial we now hypothesize for Anaximander? The earliest surviving Greek sundials date to Hellenistic times and are either hemispherical or conical. These sundial shapes make sense if the designer is imagining a spherical earth and a domed-shaped heaven, but not for Anaximander, who imagined the shape of the earth as a column drum and the cosmos as a great tree.[16] Are there any ancient artifacts that offer a clue? There is an artifact, usually taken to be a sundial but whose identity is uncertain, dating to the first century BCE and found in Qumran, along with the Dead Sea Scrolls (fig. 6.5). The scientific interpretation of this "sundial" has been the subject of considerable debate, but the conjecture here is that Anaximander's "sundial" shared a similar appearance.[17]

Having already plotted the "local noon" shadows cast by a vertical gnomon, let us take those distances as radii and make concentric circles with the

FIGURE 6.5. Qumran sundial with gnomon placed in the center.

gnomon at the center (i.e., the gnomon hole would have been round like the *empolia* in archaic column drums from Didyma[18]): the smaller circle was made by taking as a radius the distance from the gnomon to the shortest of the short shadow markers (equivalent to the summer solstice marker), and the second was made using the radius from the gnomon to the longest of the short shadow markers (equivalent to the winter solstice marker). The "frame" of Anaximander's map might very well have consisted of these concentric circles plus the rectangular dimensions emphasized later by Aristotle. Anaximander's sundial, then, looks much like a column-drum face prepared by archaic architects but, perhaps, without *anathyrôsis*,[19] unless one wishes to preserve "Ocean" running around the outside of Anaximander's map, a tradition that Ionian cartographers embraced, according to Herodotus. Indeed, Herodotus ridiculed the early mapmakers who made Ocean running round the circumference *apo tournou*, as if it had been "turned on a lathe" (or "traced with a compass").[20] We know from archaeological reports that archaic drums in Samos *were* turned on lathes, and it is perfectly believable that not only the sundial circumference, but also the inner concentric lines were turned on a lathe.[21] Thus far, the hypothesis proposed is that Anaximander's sundial resembled closely a prepared column drum, with at least two concentric circles to represent the solstices, and a third between them (but not halfway) marking the equinox. The original sundial might even have displayed *anathyrôsis*/Ocean. Indeed, the winter solstice circle might be coincidental with the inside circle of the *anathyrôsis*/Ocean band. There may also have been quite a number of sundials inspired by Anaximander, and some of them might have displayed only markers for solstices, while others also had the equinox marking, and even hours. Some versions may have had only two or three concentric circles (fig. 6.6), and thus it seems quite believable that archaeologists might have missed identifying Anaximander's sundial because it would have closely resembled a prepared column drum; of course, some examples might have been much smaller.

Next, let us consider the reports that Anaximander "discovered" or "introduced" the gnomon, according to Diogenes Laertius and the *Suda*. Scholarly opinion has dismissed the reports on the grounds that we have evidence for the use of the gnomon much earlier by the Babylonians and perhaps also by the Egyptians.[22] But let us try to see these doxographical reports in a different light. Anaximander did *something* with a gnomon that was "original." Can we make any sense of this? The report in Diogenes Laertius is that Anaximander set up his gnomon "*epi tôn skiothêrôn*" in Sparta. I take this to suggest that there was a place in Sparta where other "sundials" (shadow-catchers) were already set up,[23] and almost certainly they indicated some sort of "hour markers." And the proposal here is that Anaximander's innovation was to mark the *solstices* and the *equinox(es)* and from those markings to "frame" his map.

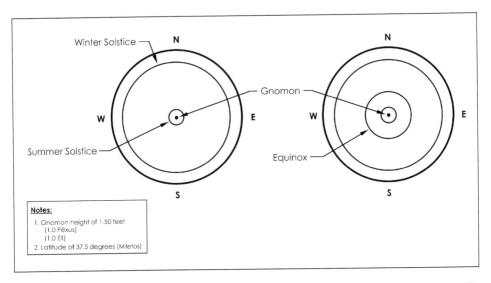

FIGURE 6.6. Anaximander's sundial with two concentric circles, summer and winter solstices (l). Anaximander's sundial with three concentric circles, summer, winter solstices, equinox (r).

To understand how Anaximander's "gnomon" was somehow *original*, let us turn to some considerations about its length. Since the numbers 9 and 10 had special meaning for Anaximander,[24] it seemed like a promising starting place to consider that his gnomon was 9 or 10 feet (or ells) in length. When shadow lengths were calculated, both 9 and 10 feet or ells could be ruled out since Anaximander would have needed a "field"—literally a horizontal surface more than 30 feet in diameter to contain the measurements of the winter solstice shadow. It is of course possible that Anaximander's "sundial face" was an "open field," but it seemed increasingly doubtful for a few reasons. First, it is almost impossible to imagine how a vertical rod 9 feet or taller could remain exactly upright in windy Greece throughout the course of the year; second, since the working hypothesis now under consideration was that the sundial face had con- centric circles, it would have been extremely difficult to imagine how to pro- duce them precisely on a 30–foot field, and third, because it seemed plausible to imagine the sundial as part of a model, perhaps the earliest example of what we have come to call a planetarium, a gigantic field would have made it near impossible to confirm the accuracy of the device.[25] So, had the seasonal sundial remained out of doors during the course of the year, the gnomon would have had to be *much* shorter; moreover, if the sundial was also connected to the map and model, a much smaller size gnomon would have been necessary.

But let us consider an objection before proceeding, namely, that a short gnomon makes it difficult to carefully measure local noon and a longer gnomon would facilitate this measurement. Let me repeat once again the caveat that this is certainly true: The longer gnomon has an important complication, namely, that in windswept Greece, it is difficult to keep a lengthy gnomon perfectly straight. A gnomon of 9 or 10 Greek feet (or ells) would be nothing short of impossible to secure from meteorological interference.

Let us continue our reconstruction comparing gnomon sizes of 1 ell, 1 foot, and .5 feet, and the resulting local noon shadows that would be cast corresponding to them for Miletos and Sparta. By modern-day measurement, with Miletos at 37.5° north latitude, and Sparta at 37.31° north latitude, we can see that the inscribed circles for summer and winter solstices would have been virtually identical. Using a gnomon of 1.5 feet (equivalent to 1 ell), the shadow length on the summer solstice in Miletos was 0.374 feet, and in Sparta using a gnomon the same size, the shadow would be 0.369 feet. Had the gnomon been 1 foot in height, the same comparative shadow lengths would have been 0.249 feet in Miletos and 0.246 feet in Sparta (fig. 6.7). And if Anaximander had used a smaller gnomon yet, let us say of .5 feet, the difference of marking the summer solstice radii (0.125 feet in Miletos and 0.123 feet in Sparta) would have amounted to a mere two thousands of an inch (0.024 inches)!

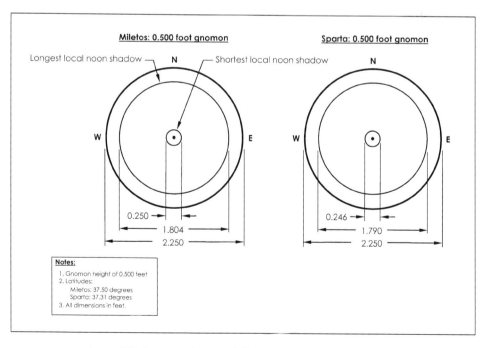

FIGURE 6.7. Anaximander's sundial comparative: Miletos and Sparta.

Of course, the gnomon would need to have been very slender and have had a very sharp point to deliver these exacting results, and the shadow markers would have had to be made—or confirmed—by the most careful observations. And had the gnomon height been *shorter* still, the inscribed circles identifying the seasonal markers would have differed from Miletos to Sparta by even less. Now, if it is neither practical nor likely that such refined measurements could be produced on a stone dial face—given the limits on sharpness of the gnomon point and the width of the inscribed circle line probably made by means of a lathe—then for all practical purposes the circles inscribed on column drums in Miletos and Sparta for these gnomon sizes—of 1 ell, 1 foot, and most especially .5 foot—would have been virtually identical. Thus, it is plausible to propose that this is what the reports of an "original" gnomon mean: Anaximander *brought with him to Sparta his own gnomon* whose seasonal markers he had already measured in Miletos. Perhaps he also brought with him to Sparta measured strings to make exactly the radii of the concentric circles that he determined in Miletos. The use of measured string or rope is the usual measuring technique of architects when, with great precision, they laid out the building plan of the great Ionic temples.[26] Thus, Anaximander could create not only a "replica" of his Milesian research in Sparta, but moreover he could *test* his understanding and, of course, have others marvel at his "scientific predictions." Understood in this light, Anaximander's gnomon was yet another display of the principle of One over Many, and this one born from, or at least confirmed by, careful observations, not *muthoi*. Thus, the finished appearance of Anaximander's sundial (without hour markers), then, is more or less the *same* as a prepared column-drum face—both exhibit concentric circles on a flat disk. If this thesis is correct, the archaeologist may yet find Anaximander's sundial. Its general appearance is presented in figure 6.8, contrasting the results with a 1-foot gnomon versus a gnomon 1 ell (1.5 feet) in height.

Let us now follow through the consequences of this proposed interpretation of the doxographical reports. Since it is entirely believable that Anaximander visited Naucratis since the Milesians alone of the Greeks were granted the privilege of establishing a colony in Egypt[27] (or even if he did not, he could have sent *his own gnomon* with a compatriot), if his sundial experiment was repeated in the Nile Delta with the *same* gnomon (and the *same* string lengths that he verified for both solstices and equinoxes in Miletos and Sparta), he would have realized that the measurements were significantly *different* at this more southerly location. In Naucratis, the markings for the summer and winter solstices, and hence the concentric circles, have much *smaller* radii (fig. 6.9). Hence, Anaximander would have been aware that the circles appear much closer to the gnomon as he traveled south. What did this result plausibly mean to Anaximander? Well, unlike the uniform results he experienced when he traveled east and west, from Miletos to Sparta, the results

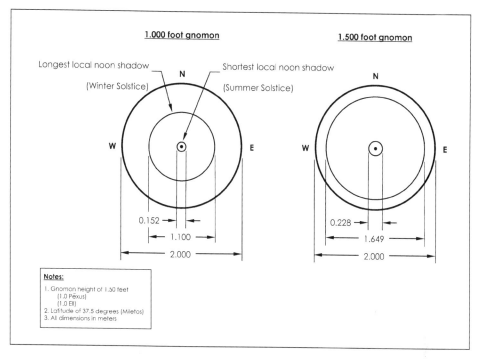

FIGURE 6.8. Anaximander's sundial with diameter measurements for 1-foot and 1.5-foot gnomons in Miletos.

were quite different as he traveled south and north—*although he brought with him the same gnomon!* There is a report that Anaximander founded a colony, Apollonia, on the Black Sea.[28] There can be some doubt about his founding a colony because, had he done so, we should have expected him to be known as "*Anaximandros tou Apolloniou*" and not "*Anaximandros tou Miletou*" as he is known.[29] Nonetheless, the report allows us to consider further that he was a much-traveled man and may well have ventured northward to the Black Sea region. Had he made measurements there with *the same gnomon*, he would have also seen that the summer solstice circle is significantly *larger* than at Miletos, while as he traveled south the concentric circles would become increasingly *smaller*. And so it is plausible to conclude that Anaximander believed that the center of the flat-disk earth was *south* of Naucratis, and indeed south of Heliopolis.[30] If this hypothesis proves correct, then Anaximander's map would have identified ancient Cyene (modern-day Aswan), not Delphi the *omphalos*, as the center of the earth. The key reason that justifies this conclusion is that when Anaximander realized that the summer solstice circle got smaller as he (or his compatriots taking his gnomon to make

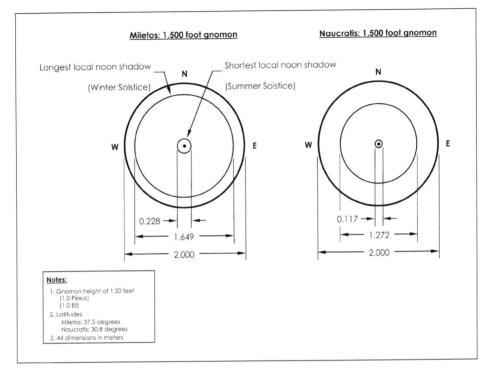

FIGURE 6.9. Anaximander's sundial comparative: Miletos and Naucratis (emphasizing smaller solstice circle diameters in Naucratis).

measurements) traveled *south*, he could have inferred that he was moving *closer to the center* of the flat-disk earth, and would get closer and closer still had he ventured further south from Naucratis in Egypt (fig. 6.10). Thus, he grasped that *his* location on the earth, revealed by the outlines of the "earthly map" that in fact appeared by the shadows on the column-drum sundial face, was always *on* the circle marking the summer solstice—indeed, on the *north part of the circle*. Thus, when he reached the center of his projected earthly map, there would be *no* summer shadow cast on the summer solstice, and thus he would be *at the center* (*en meso*). Thus, Anaximander would have realized that *his* location was always on the (northern or upper part of the) smallest concentric circle on the drum face that he inscribed in making his map, and furthermore that Miletos, Sparta, and Delphi were significantly north of center but located between the concentric circles marking the summer and winter solstices. And since all map makers, ancient or modern, have to make an arbitrary judgment about where to set the "center" in making their map, Anaximander may well have selected what

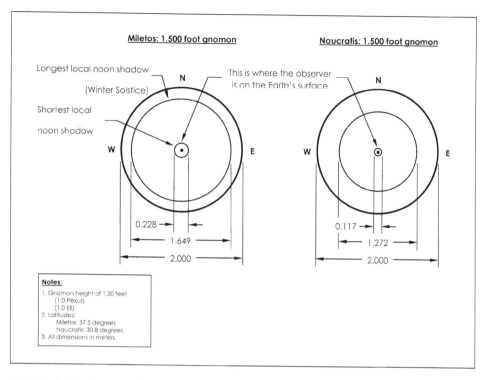

FIGURE 6.10. Anaximander's sundial comparative: Miletos and Naucratis (emphasizing the observer's location on the earth's surface).

we call nowadays the Tropic of Cancer, 23$^{1}/_{2}$ degrees north of the Equator *based on the results he collected from his sundial experiment.* Of course, since he imagined the earth as a flat disk, there could be no meaningful identification of "latitude" or "equator" or "tropics" as such.

While it is true that these opinions are very strange for a Greek in the archaic period, we need only recall that Anaximander held other startlingly nontraditional views. He was the first of the ancients, according to Aristotle,[31] to hold that the earth remained aloft at the center (*en meso*) of the cosmos *held up by nothing*; he maintained, according to Hippolytus,[32] against the view of a crystalline sphere where all celestial objects are fixed at exactly the same distance from us, that there was such depth in space that the sun was immensely farther from earth than the immensely distanced moon; and from Pseudo-Plutarch[33] we learn that Anaximander claimed that humans were descended from some kind of fish: first that we began *in* the water, and, acknowledging the helplessness of infants, we must have been in a protected environment much longer to guarantee our survival to the present day. If we keep in mind how "nontraditional"—shall we say revolutionary?—were Anaximander's ideas, the possibility that he held that southern Egypt was the center of the flat-disk earth is much less untenable.

The reports from Diogenes Laertius and the *Suda* placed together the sundial, the map, and a *sphairos*. The term "*sphairos*" refers plausibly to a "model" of some sort; it could not be a "sphere" because Anaximander's cosmos was not spherical—the cosmos was cylindrical like a great tree. So, perhaps *they were all one thing* set up in Sparta. Had the gnomon been 1 *pous*, Anaximander would have needed a column drum of roughly 4 feet in diameter to contain both solstice circles and still have Ocean running around the circumference; had he selected instead a gnomon of a *pexus* (1 1/2 feet = 1 ell), he would have needed a column drum of roughly 6 feet (i.e., 2 meters).[34] This idea of a column drum, perhaps in the proportion of 3:1 with a diameter of roughly 6 Greek feet (or 4 ells) seems like a particularly attractive hypothesis because such a size would have been more or less identical to the architect's module, judging by the Ionic temples in Didyma, Ephesos, and Samos.[35] However, had Anaximander's gnomon been .5 feet, the dial face needed to be only 2 feet in diameter, a size that could have been even more easily turned on a potter's wheel. And in all three assignments of gnomon lengths, the comparative results at Miletos and Sparta would have been more or less indistinguishable, especially for the smaller sizes.

Diogenes' report is that Anaximander set up his gnomon in Sparta.[36] The proposed design—resembling a column drum—might have been seated on a slender stand, perhaps resembling a column, but much narrower than the drum itself. It is likely that the drum face had three concentric circles marking the summer solstice, winter solstice, and the equinox(es). Traveling from Miletos, he would have known beforehand the exact distances—the radii—to each of the concentric circles, by means of previously measured strings. Of course, the sundial would have been testable in front of the whole community. Moreover, the circumference of the drum would have contained markings for the rising and setting of the sun on the solstices and equinox, and on the drum face a "map" of the earth might well have been inscribed that had at its center what we now call Aswan, Egypt, or in any case south of Naucratis, and certainly far to the south of Delphi. Finally—the *sphairos*—we can imagine that two "wheels," possibly made of bronze, representing the path of the sun, could have been attached (at the drum sides), one depicting the sun at summer solstice and the other the sun at winter solstice, and both wheels would have been continued *under* the earth, one of Anaximander's truly awesome cosmic speculations. By supposing that the column-drum sundial would have been supported by a slender column, he could show how the sun wheels continue under the earth. These bronze "wheels," now hypothesized, would certainly not have been made to scale, since the sun-wheel orbit is at least 54 earth-diameters.[37] There would have been no way to make such gigantic wheels unaffected by wind; had they been made to scale, the sundial would have been so tiny that reading the shadows on the dial face would have been quite impossible. So, these bronze wheels now hypothesized, while not made to scale, would have conveyed that the sun wheel goes *under* the earth and the earth stands in the middle. This, then, was the model, the map, and the sundial, made possible by the use of a uniform gnomon whose shadow-casting properties had been confirmed by trial and error, all mentioned together but not connected explicitly in these very late reports. Figure 6.11 is a reconstruction of the proposed model "set up" in Sparta.

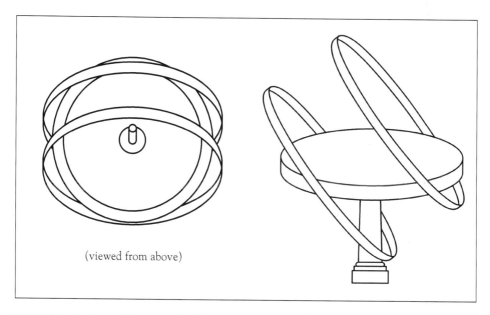

(viewed from above)

FIGURE 6.11. Scale model (*sphairos?*) of Anaximander's sundial with summer and winter solstice "sun wheels."

Now the time has come to consider the problem of the equinox. Scholarly opinion has been doubtful about Anaximander's measurement of the equinox, and the reason commonly proposed is this: A knowledge of the obliquity of the ecliptic is required to correctly identify the time of the equinox, and that discovery belongs not earlier than Oinopides of the fifth century BCE.[38] The "obliquity of the ecliptic" is the angle between the planes of the ecliptic and the celestial equator. On the celestial sphere, it is the angle at which the ecliptic intersects the celestial equator. Now the ecliptic is a great circle on the celestial sphere that represents the apparent path of the sun in its motion relative to the background stars. This great circle is known as the "ecliptic" because eclipses can occur when the moon crosses it. The celestial equator is the great circle on the celestial sphere obtained by the intersection with the sphere of the plane of the earth's equator. When the sun lies in the plane of the celestial equator, day and night are everywhere of equal length; this occurs exactly and only two times each year. The "equinox" is the instant at which the sun crosses the celestial equator; the sun is then vertically overhead at the equator, and day and night have equal duration at every point on the earth's surface. The apparent annual path of the sun on the celestial sphere is inclined to the celestial equator and intersects it at two points. The terms "vernal equinox" and "autumnal equinox" are applied to these points (fig. 6.12).

Thus, the objection has been that Anaximander could not have identified the equinox because he could not have known the obliquity of the ecliptic. This objection proves to be rather beside the point. For we need to imagine the astronomical picture that developed from Anaximander's approach to the gnomon and circular but flat-faced sundial, reflecting his conception of

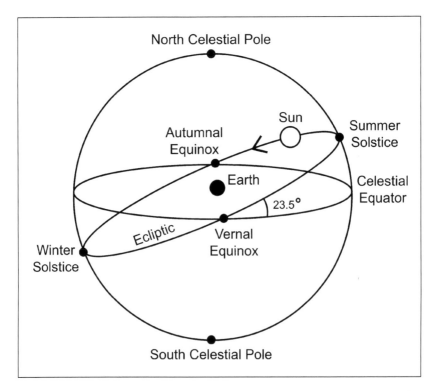

FIGURE 6.12. The celestial sphere and the obliquity of the ecliptic: identifying the exact moments of the equinoxes.

the earth as a column-drum–shaped cylinder and not a sphere. By means of the sundial construction that has been proposed, Anaximander could have identified the equinox to the day, but the case that he could have known it to the minute cannot be established. Let us enumerate the argument.

We have now explored the hypothesis about how Anaximander made his sundial, how he plotted, or confirmed, the shortest shadows each day and made concentric circles to identify the solstices, and how he placed marks on the circumference of his sundial (microcosmically mimicking the markings obtainable from the column-drum earth itself) to indicate the rising and setting of the sun each day through the course of the year. The mark for the sun at its rising farthest north of east defines the day of the summer solstice; the mark of the sun at its rising farthest south of east defines the day of the winter solstice. Now, in order to identify *the day of the equinox*, Anaximander had only to *bisect* the angle formed by the lines connecting the rising (or setting) of the sun on summer and winter solstices; it was not necessary either to have an understanding of the ecliptic or the capacity to calculate the equinox arithmetically. Bisecting the angle merely confirmed his construction at the outset on the sundial face. At that earlier stage, according to our hypothesis, Anaximander made a line perpendicular to the north–south line of the shortest shadow markers of the day throughout the course of the year. The line perpendicular to this north–south line not only makes a diameter that bisected the circular-drum face but it at once identified the cardinal direction of due east–west, the rising and setting points of the sun on the equinox itself. Bisecting an angle in a circle was simple and straightforward in the tradition of archaic architects, who routinely worked the geometry of the drum faces they were preparing for installation in the colonnades of the temples.[39] If Anaximander needed a first or second opinion on bisecting an angle, he could have consulted the architects working in his own backyard or just watched them work. Thus, as he sited the sun rising (or setting) on the east–west line, and confirmed by the short shadow mark at local noon, he would have known the exact day *when* the equinox would take place (fig. 6.13). On that day, identifying the shortest shadow mark, Anaximander would have been able to produce the third, intermediary concentric circle by means of that radii measurement. It is important to note, therefore, that Anaximander's identification of the equinox was a result of simple but careful observation, not computation, though he might afterwards have come up with an arithmetical formula counting the days of increasing (or decreasing) daylight and confirmed the equinox point.

According to two doxographical reports, Anaximander's sundial also had "hour indicators."[40] Of course this is possible, but we must first get clearer about what kind of "hours" these could be. Anaximander might have been familiar with the same sources known later to Herodotus that claimed the Babylonians had divided the day into twelve hours.[41] And if he

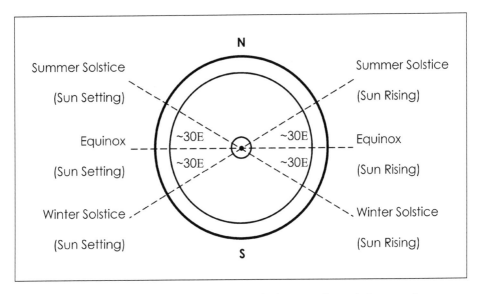

FIGURE 6.13. Anaximander discovered the equinox by bisecting the angle between the summer and winter solstice risings, and the summer and winter solstice settings.

proceeded with this view in mind, he approached the sundial face, taking as his starting line the diameter made by the east–west line, and the north–south line perpendicular to it formed by the shortest shadow markers, thereby creating a right angle. He could have bisected each angle a series of times to achieve the "twelve parts." Without committing to the number of "hour markers" that Anaximander's sundial displayed, it seems most likely that Anaximander's technique of "hour constructions" consisted in bisecting a series of angles (fig. 6.14).

Thus, had the sundial also displayed hour indicators, additional radiating lines would have appeared. This result could have been achieved in a straightforward way by the simple geometrical technique of bisecting each angle—that is, by the same technique by which he discovered geometrically the equinox.

In arguing that Anaximander identified the "equinox," there are two key points: (1) the baseline of a sundial displaying hour markings is a diameter bisecting the circle running due east–west, and that line itself identifies the day of the equinox, and (2) calendar reform in Miletos (and elsewhere in Greece) depended on the identification of the equinox—the vernal equinox in the case of Miletos and perhaps the autumnal equinox elsewhere in Greece,[42] where the New Year was announced by the new moon following that seasonal marker. The report that places Anaximander in Sparta setting up a *seasonal* sundial suggests that he recognized that the

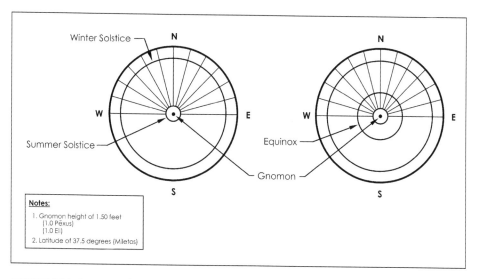

FIGURE 6.14. Anaximander's sundial with two concentric circles, summer and winter solstices, and hour markers (l). Anaximander's sundial with three concentric circles, summer, winter solstices, equinox, and hour markers (r).

one device he invented in Miletos, and installed in Sparta, could resolve comparable problems of the solar calendar and a fortiori aid in calendar reform throughout Greece.[43]

D

OBJECTING ARGUMENTS AND SUMMARY

There are two different but formidable objections that deserve to be raised against these hypotheses in general and the line of reasoning advocated here. And now is the time to explore them. Let us be clear that we must keep in mind that despite Anaximander's bold and brilliant insights, he was mistaken in important ways about the size and shape of the earth and the distances to the stars, moon, and sun. He was wrong, and so important observational inconsistencies—and hence objections to his conceptualizations—will appear as a matter of course. The underlying problem is to try to get clearer about how he thought about these results, whether he was even aware of certain inconsistencies, and what concerns were genuinely raised by his methods.

To understand the gravity of the objections we will now consider, let us first review the hypotheses being explored. At all events, our hypotheses presume that Anaximander made use of a variety of architectural techniques but

not that he used them exclusively, consistently, or at every available opportunity, and that, like architects, he made careful measurements and confirmed some of his claims by observations. Anaximander's seasonal sundial
made use of a gnomon placed vertically, not horizontally. By means of this
gnomon, Anaximander had determined in Miletos the length of the shortest
and longest local noon shadows—the summer and winter solstice markers—
and measured them by two pieces of string just as architects routinely measured out their buildings. When he traveled to Sparta and set up the *same*
gnomon, and measured out the short shadow markers with the *same* pieces of
cord, he discovered that the summer and winter solstice markers were, for all
practical purposes, identical. But when he or one of his compatriots traveled
south to Naucratis (or, for that matter, north to Apollonia on the Black Sea)
with the same gnomon and measured strings, the results were neither com-
patible nor analogous. The disparities between the measurements north and
south, as opposed to east and west, likely had far-reaching consequences for
Anaximander's map.

The first of two important objections is that Anaximander did not
engage in careful observation. Cornford had argued long ago that there was
no sign of observational experimentation in Anaximander's achievements.[44]
More recently, Couprie championed the same general theme with regard to
the seasonal sundial. Couprie had proposed a purely geometrical construction
of the seasonal sundial,[45] that is, a sundial confirmed only roughly by observations. He had argued in effect that Anaximander and his contemporaries
realized that during about one hour around local noon it is very hard to see
any differences in the length of the gnomon's shadow, especially when you
have a rather short gnomon. Therefore, Anaximander certainly used another
method to construct his sundial, one that provides a rather accurate result
and that can be achieved in one day, so that no years of careful observation
are needed.[46] According to Couprie, the best result can be achieved by making measurements both in the early morning and in the late evening, when
the shadow of the gnomon is very large and changes rather quickly in length,
and, moreover, the taller the gnomon, the better the results. Couprie
describes the procedure, more precisely, as follows (see fig. 6.15): mark the tip
(A) of the shadow of the gnomon at some (early) time in the morning. Then
draw through that point the arc of a circle with the base of the gnomon (G)
as its center. Then mark in the afternoon (late) when the point of the shadow
(B) hits the arc. Finally, bisect the angle AGB, and the line found is
the north–south line.

It is not implausible that Anaximander, at some point, imagined the
construction of a seasonal sundial by means of such a geometrical technique.
In support of this view, I had already argued in an earlier publication that
sixth-century architects routinely *trisected* the surface of the column drum as
part of the technique to ensure that the drum—some six feet in diameter—

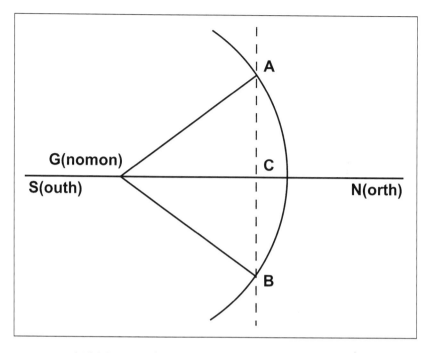

FIGURE 6.15. Measuring the equinox by means of a geometrical construction.

was level throughout the surface. And this of course means that the practical technique of bisecting and trisecting a circle was understood contemporaneously. Figure 6.16 is Hellner's illustration of the technique of trisecting the circular drum surface.[47]

Hellner's drawing illustrates the geometrical technique of trisecting a circle. First, the architect bisects the circle. Then, given that diameter, an arc the length of the radius sweeps though the center with the compass point placed perpendicular to it. The two places where the arc intersects the circle are the trisecting points (fig. 6.17).[48]

The evidence for this practical skill in applied geometry is found on the column drums at Samian Dipterous II, the temple to Hera dating to the second half of the sixth century, seen in figure 6.18.[49]

The evidence shows that trisecting a circle was familiar to architects working at the temple building site, and thus the kind of geometrical technique supposed by Couprie is entirely believable. But how would Anaximander have known that the technique was successful? Had he initially made his construction by this geometrical method, the success of his method could have been shown only by confirming the results *observationally*. To do so, Anaximander would have had to check the markings for summer and winter

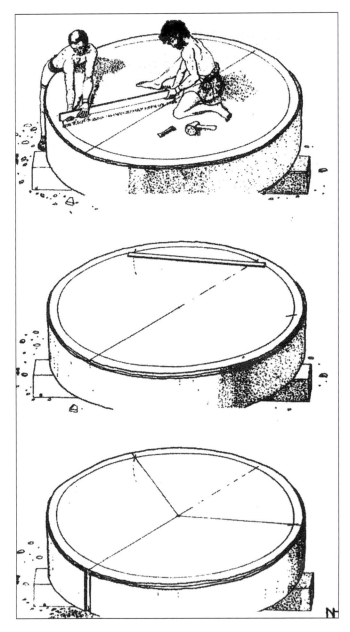

FIGURE 6.16. Trisecting a column-drum face (i.e., a circle), illustration by Nils Hellner.

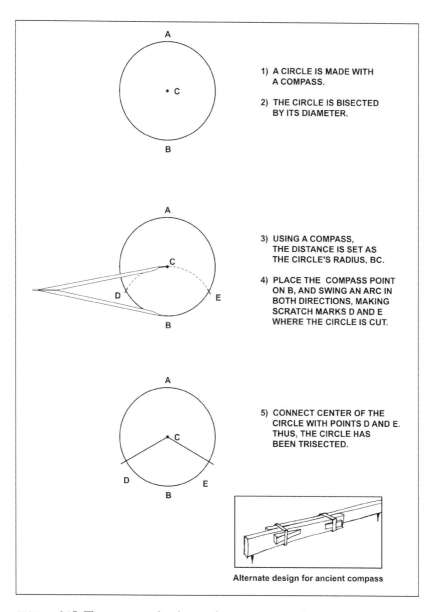

FIGURE 6.17. The geometrical technique for trisecting a circle, or column-drum face.

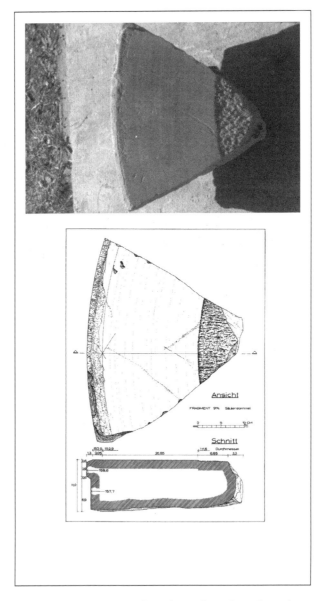

FIGURE 6.18. Fragment of a column drum from the archaic temple to Hera in Samos, Dipteros II, exhibiting markings for trisecting the drum face.

solstice by observing, for instance, whether the sun ever rose north of the summer solstice marker that he would have indicated and whether it ever rose south of the winter solstice marker. To check this, he would have needed to keep a dial in place throughout the course of the year, and to determine by observation—by sightings—the straight line that connected the gnomon with the summer and winter solstice markings, just as architects determined their straight line by means of a marking system.[50] When Anaximander set up his seasonal sundial in Sparta, he would have been sure that the accuracy of his instrument would have been confirmed or refuted, celebrated or disclaimed, by the Spartans themselves. Even if Couprie were right in claiming that Anaximander first set up his sundial as a geometrical construction only, the proof of his markings required careful observation and measurements, by him and his compatriots, to *confirm* his success. Moreover, Couprie never considered that the dial itself would have been a column drum or a drumshaped stone, even though the geometrical construction he proposed required an arc that could certainly have been indicated by the circumference of a drum.

Moreover, Couprie's method supposes the use of a *tall* gnomon, not the short one proposed here, because the markings are easier to distinguish, especially approaching local noon. But this proposal is suitable only for a single day or two when clear and calm weather prevails. A tall gnomon requires a very substantial open space to measure the results and would be entirely impractical in Greece where, through the course of the year, the gusty winds would prevent the gnomon from remaining precisely in the same place. It is difficult to imagine how it could have been set in place during the course of the whole year had it been tall, and, if it did not remain in place, it could hardly have served as a seasonal sundial. Had the gnomon been ten feet tall, as we have already considered, Anaximander would have needed a field some thirty feet long to prove the accuracy of his identification of the winter solstice! And the tall gnomon would have had to stand in an open, unprotected area throughout the course of the year to exhibit the results. This scenario is implausible. Thus, for this first objection, even if Anaximander had started his sundial construction by means of a geometrical technique, careful observations would have been required to confirm the success of this practical instrument. The importance of observational measurement for Anaximander has been grossly underestimated.

The second important objection emerges when the results east and west—Miletos and Sparta—were compared with those made north and south—Black Sea Apollonia and Naucratis. What were those consequences for Anaximander's cosmic imagination, keeping in mind that Anaximander had shared the view of both Homer and Hesiod that the earth was flat?

Anaximander had most certainly paid attention to the changing elevation of the sun as the first day of summer approached, when the sun would be highest in the sky. Had he measured the sun's angle of inclination at local noon on the summer solstice when the sun cast the shortest shadow both of the day and of the year, he would have discovered that the measured angle in Miletos was more or less 76°. The hypothesis we have been exploring requires that Anaximander would have known that as summer approaches as one travels southward from Miletos toward Naucratis, the sun reaches a higher apex, and, in fact, in southern Egypt (in ancient Cyene, modern-day Aswan) is directly overhead with its zenith at 90° on the first day of summer. But a consequence of Anaximander's cosmic numbers—the distance to the sun, the farthermost wheel in the heavens, is 27 earth-diameters—is that it seems to undermine this very claim. Why and how?

I have been at pains to argue that Anaximander's cosmic numbers have a symbolic meaning and are not the result of any observable techniques or testing. Moreover, that while Anaximander had made use of architectural techniques and technologies because he imagined the cosmos as cosmic architecture, he seems to have employed them selectively and adapts them for his own distinct purposes. His selection of cosmic numbers suggests distances by means of an archaic formula embraced by Homer and Hesiod and others of that age that, in the absence of adequate observable techniques, those numbers mean to convey distances as *far*, *farther*, and *farthest*, while at once making the remarkable point that there are unfathomable depths in space itself. If the distance of the sun is treated like the observational claims of the shortest local noon shadow in the year, or the rising of the sun against the horizon most north-of-east, thus identifying the summer solstice, the result is to undermine the hypothesis we have been exploring. This is because when Anaximander adopts the view that we are living on a flat earth 27 earth-diameters distant from the sun, the sun must be both very close and comparatively small (see fig. 6.19). Had Anaximander taken the cosmic numbers as "observationally" significant as he must have regarded shadow measurements in his sundial experiments, he would have realized that this commitment made it impossible for the sun to be overhead anywhere at all on a flat earth.[51]

The geometry of the cosmos, then, where the sun stands at a distance of 27 earth-diameters is at odds with the experience that the sun is, in fact, higher in the sky as one travels southward from Miletos. Was Anaximander aware of this? Did it matter to him? We know that the Milesians had a colony in Naucratis and consequently they knew very well that it was *much hotter* in Egypt; this simple observation coincided with the simple observable fact that the sun was *higher* in the sky (than it was in Miletos) as summer approached, indeed higher on any given day. So, which alternative do we embrace and which do we discard? Which is easier to reject? Do we give up the hypothesis of the vertically set gnomon, and the testable, confirmable measurements of the shortest and longest local noon shadows in Miletos and Sparta, over

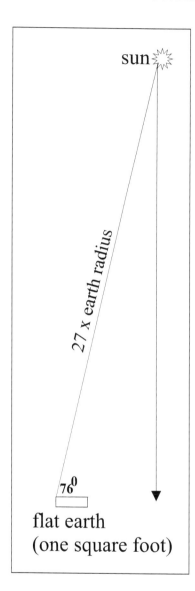

sun

27 x earth radius

76⁰

flat earth
(one square foot)

FIGURE 6.19. If the sun is observed at an altitude of 76° in Miletos at local noon on the summer solstice, then the point where the sun appears directly overhead is well beyond the surface of the flat earth.

and against those in Naucratis (and Apollonia)? Or do we reject the supposition that Anaximander treated the numbers to describe the cosmic distances as observationally appropriate and useful? It seems clear that we must reject the latter and preserve the former; we must reject the claim as observationally significant that the sun's real distance is 27 earth-diameters. Anaximander regarded the formulaic distances to stars, moon, and sun (9 + 1 earth-diameters, + 9 to the moon, and + 9 to the sun) as poetically but not observationally meaningful. It makes more sense to suppose either that he never worked out (or realized) the observational inconsistencies of maintaining a 27 earth-diameter cosmos, or that he never thought the archaic formulas to express cosmic distances were to be treated like the testable, confirmable shadow-casting properties of a gnomon in his seasonal sundial. He never regarded those cosmic numbers to represent real distances anymore than Hesiod imagined the real distance from the highest heaven to the earth was accurately expressible by a falling anvil. Let us review this cosmic geometry, one more time, in light of Anaximander's suppositions about a *flat* earth. In this review, we emphasize the inconsistencies that follow from Anaximander's cosmic conceptions when recounted as a consequence of maintaining that the earth is flat.

In light of today's knowledge, we commonly hold the view that the sky (with an unobstructed horizon) observed from a large sphere (such as earth) appears hemispherical with an angle of 180°. Now if we begin by supposing that the sun is *extremely distant*, we might think of the horizon as a plane that is tangent to the sphere at the observer's position. As an observer moves on the surface of the earth, the horizon moves as well (i.e., it essentially rotates about the center of the earth with the tangent point underneath the observer). Thus, the altitude of Polaris or the sun (on a particular day) varies with the latitude of the observer, and the time for an astronomical event (i.e., local noon) varies with the longitude between two positions. The evidence we have about Anaximander's cosmic picture does not suggest he was aware of these observational complexities.

Now when we consider Anaxamander's cosmic geometry, the column-drum earth is flat, so the "horizon plane" previously described must be fixed. Therefore, movement on the surface of the "flat earth" *should* result in no change in the observer's horizon. If an observer's horizon does not change, then the altitude of an astronomical object (i.e., at a nearly infinite distance) is minute, and would not be observable. This means that the sun's angle would differ by only 0.005° between observers on the north and south "poles" of a flat earth. Thus, the altitude of the sun on a particular day (i.e., winter solstice) would be essentially the same for all observers on the "flat earth." And this also means that the angle of Polaris above the horizon would be essentially the same everywhere on the earth; in fact, Polaris *should* be visible to all earthly observers. This geometrical conception cannot square with the simple observation of the North Star in both Miletos and Naucratis; in Naucratis, the North Star must be observed lower in the sky. And we have already acknowledged that the Milesians had a colony in Egypt; surely, they must have known that the horizon had changed visually from Miletos. How, then, should we understand Anaximander's cosmic claims, given his supposition of a flat earth?

And there is yet another difficulty to be considered along with these problems: The change in altitude of the sun on the same day as witnessed by observers at different positions on the flat earth increases as the distance to the sun decreases. However, this change (distinct from the earlier case where the sun is supposed to be very distant) is not sufficient to explain the "true" altitude of the sun in Anaxamander's model. Anaxamander's cosmology places the sun at a finite distance of 27 earth-diameters from the "flat earth"—and this means "extremely close." If the sun is observed at an altitude of 76° in Miletus at local noon on the summer solstice, then the point where the sun appears directly overhead is well beyond the surface of the flat earth. Consequently, constrained by Anaximander's cosmic numbers, the sun should never be able to appear directly overhead on the summer solstice as observed from the "flat earth," but it did, as a matter of fact, appear just this way in southern Egypt. So, how do we square this observation with his implied cosmic geometry?

Let us consider this problem in terms of one more illustration. Imagine yourself standing on a 1 foot by 1 foot tile in the corner of a room (i.e., 1 square foot flat earth).[52] Next, imagine the sun 27 feet away, and at an altitude of 76° (i.e., the opposite corner of the room). Now imagine yourself in Miletos—on the 1 foot by 1 foot tile—on the summer solstice. In order for the sun to appear directly overhead, you would have to walk to the opposite corner of the room. However, the "flat earth" in this scale is only 1 square foot, so the sun could never appear directly overhead as observed from the earth. Furthermore, as you move your eye horizontally around your 1 square foot (flat earth) surface, the sun doesn't move much at all.

The objecting arguments seem to place Anaximander in one of two camps: either he was unaware of inconsistencies or they did not trouble him. Thus, either he was unaware that the summer sun in Egypt was significantly higher in the sky than in Miletos—which is almost impossible to accept—or he was unaware that its altitude significantly in excess of 76° was inconsistent with his other suppositions about the flat earth and the comparative closeness of the sun (at 27 earth-diameters) required by his cosmic numbers. Is there, perhaps, another resolution to the apparent inconsistencies? There is another response that offers the prospect of explaining, but only partially, how Anaximander might have known that the visual experience of the horizon changed—the sun's altitude and night sky—while traveling south to Egypt and still maintaining that the earth was flat. It will not, however, resolve the problems of identifying the distance to the sun with 27 earth-diameters; this problem will only be defused by accepting that the archaic and poetic formula had no observational bearing for him. Perhaps Anaximander held the same kind of view that was later expressed by Democritus, according to the report of Diogenes Laertius,[53] namely, that *there is a sharp titling of the earth toward the south*. While this passage occurs with specific reference to explaining eclipses,

"Most scholars have assumed a gap [in the fragment] so that some other phenomena than eclipse is explained by the tilting of the earth."[54] The supposition of a sharp dropping off of the earth toward the south might have reassured Anaximander that the visual changes in the sky did not undermine his supposition of a "flat-ish" earth. Thus, the sun could "appear" higher in the sky in Egypt, a fact that he almost certainly would have known, without requiring him to abandon the claim of a flat earth.

No one who considers the archaic tradition of early mapmaking, or the archaic tradition that identifies Delphi as the *omphalos* of the world, can escape the stunning reality that either the cosmos is inclined 23 1/2° north or that the earth is inclined 23 1/2° north. How shall it be explained why "north" is not directly overhead? Might part of the explanation consist in just this: The Greeks held that there was a falling away of the earth to the south, and this accounted for the observational discrepancies—for Homer, Hesiod, and the archaic Greeks who held that the earth was a flatish disk—in the sightings of the altitude of heavenly appearances on the changing horizon, the sun, and in the night sky? The problem that the sun is directly overhead in southern Egypt, although his cosmic distance of 27 earth diameters to it makes this impossible, can be lessened not by these considerations but only by appeal to the likelihood that he did not regard the poetic formula of 9 + 1 and 9 to be observationally significant.

PART II

Archaeology and the Metaphysical Foundations of an Historical Narrative About the Origins of Philosophy

SEVEN

The Problems

Archaeology and the Origins of Philosophy

A

THE PROBLEM OF PHILOSOPHICAL
RATIONALITY AND CULTURAL CONTEXT

IN TWO RECENT STUDIES—*Anaximander and the Architects: The Contributions of Egyptian and Greek Architectural Technologies to the Origins of Greek Philosophy* (2001),[1] and again in *Anaximander in Context: New Studies on the Origins of Greek Philosophy* (2003)[2]—I made use of archaeological reports and artifacts to illuminate aspects of Anaximander's thought, that is, his rationality. In the process, however, it has become clear to me that the methodology I employed had never before been explicated and presented as a method for future research. In *Anaximander and the Architects*, I showed that among the many recognized sources that influenced Anaximander's thought, architects and their monumental temple building projects had never before been appreciated. I urged readers to see, perhaps for the first time, Anaximander and Thales in the community-of-practical-interests with the architects Theodorus and Rhoikos, Cherisphron and Metagenes, working contemporaneously in their very backyards. I appealed to a range of architectural techniques, most especially plan and elevation views, and the technique for preparing column drums for installation, called by the modern term of *anathyrôsis*, discussed by archaeologists and evidenced in their excavations, to illuminate points of view by which Anaximander imagined the cosmos *because* he came to see the cosmos in architectural terms. In my earlier work, I appealed to archaeological evidence and reports without ever precisely explicating how I was using them.

In *Anaximander in Context*, I continued this approach, trying to get clearer yet about Anaximander's cosmic numbers and proportions. Since I had argued already that Anaximander imported architectural techniques into his cosmic imagination and conceptualization, I inquired further about how the numbers and proportions he selected to describe the distances to the sun, moon, and stars might well have shared more similarities with the architects' temple architecture. Thus, by appeal to the archaeologists' reports and artifacts, I not only showed in more detail what Anaximander likely watched at the temple building sites, but I also extended the research into contemporaneous archaic sculpture by exploring what archaeologists reported on numbers and proportions used in these associated techniques as well.

But in both studies, while I made use of archaeological resources, the method by which I appealed to them was never explicated. No one before had made serious appeal to archaeological reports and artifacts to illuminate Greek philosophy; I realized that this defect was emblematic of the whole discipline of scholarship in ancient philosophy. Without exemplars, a new method needed to be proposed, and a central theme of this new study is to devise and provide, and then employ systematically, this new method. But to see why there have not been exemplars that I could follow or critique is to come to understand why scholars of ancient philosophy have not appealed to archaeology in the first place. To this consideration I turn first.

One of the leaders in the field of early Greek philosophy, Jonathan Barnes, produced a two-volume work[3] [1979, 1982] that is arguably the most influential full-length study on the pre-Socratics in the last fifty years. That work, together with many other studies that Barnes inspired, takes an approach that reflects a certain analytic conception of "rationality." Despite all its strengths, and they are many, this point of view is inadequate to help us account for the *origins* of philosophy that Aristotle traces to the Milesians. That is, Barnes' approach is not adequate to account for the kinds of thinking that appeared among the Milesians, the earliest philosophers. What is at stake in the debate I am trying to foster is a vision about the nature of philosophical rationality and, consequently, the issues and approaches that are appropriately a part of it.

In his first edition, Barnes marked out part of the frame for his map of early Greek rationality:

> I have little concern with history. It is a platitude that a thinker can be understood only against his historical background; but that, like all platitudes, is at best a half-truth, and I do not believe that a detailed knowledge of Greek history greatly enhances our comprehension of Greek philosophy. *Philosophy lives a supracelestial life beyond the confines of space and time.*[4]

Thus, in the first edition, Barnes made clear that philosophical rationality is outside the spatial and temporal borders reached by studies that take seriously matters of historical and cultural context. In the second edition, he responded to his critics who took umbrage at so extreme a view. But, indeed, he merely reaffirmed his extreme stance on the province of "philosophy" and the rationality that is its subject matter. The result is to provide a picture of philosophical rationality that dismisses matters of *cultural context* as irrelevant to such a study.

> Some critics, indeed, have accused me of being anti-historical, and their accusation has some point: I made one or two naughty remarks about history, and I occasionally flirted with anachronistic interpretations of Presocratic views. . . . In speaking slightingly of history I had two things in mind—studies of the 'background' (economic, social, political) against which the Presocratics wrote, and studies of the networks of 'influences' within which they carried on their researches. For I doubt the pertinence of such background to our understanding of early Greek thought.[5]

Thus, Barnes made clear what did and what did not properly count as the rationality that is philosophical. It is my hope that the following discussion urges us to explore, and perhaps redraw, the borders of what counts as philosophical rationality. To investigate this matter, I will now consider the overall problem that is presented by an appeal to archaeological reports and artifacts to shed light on early Greek philosophy, having placed these reflections *after* the series of case studies in Part I. If the reader comes to conclude that the chapters of Part I of this book bring new and relevant light to our understanding of Anaximander's rationality *by means of issues that rely on cultural context*, then the exclusively *a*historical and *a*cultural approaches recommended by those like Barnes deserve to be modified, if not abandoned.

While proponents of analytic philosophy, like Barnes, in effect hold that their methods are sufficient to establish what is true and false and what is reasonable to believe, and that historical inquiry is irrelevant, the weaknesses in such suppositions have been clearly articulated by MacIntyre and others. For historical inquiry is required not only to establish what a particular point of view is, but also that it is in its historical encounter that any given point of view establishes or fails to establish its rational superiority relative to its particular rivals in some specific context.[6] The very idea of an ahistorical and acultural rationality, despite the protests of its advocates, must always and ultimately be explored within an historical period or culture in order to grasp its very nature.

B

THE PROBLEM OF ARCHAEOLOGY
AND GREEK PHILOSOPHY

Archaeology has rarely, if ever, been appealed to by scholars of ancient philosophy in order to clarify or deepen our *philosophical* understanding of early

Greek philosophy. Why this is so merits pause. Whatever, exactly, the business of philosophy may be, the discipline that researches "ancient philosophy" has not even entertained the possibility that the reports and evidence from archaeologists could be philosophically illuminating, judging by the fact that there is no secondary literature involving it significantly.

Stated differently, we might observe that the study of ancient Greek philosophy is one of the few disciplines that explores the world of Greece without regarding it of disciplinary significance to actually *visit* the country itself. From the perspective of contemporary scholarship in the field, a visit to Greece might prove to be "inspiring," a source of all kinds of information about Greek cultures, both ancient and modern, and great fun, but it would not likely prove to be *philosophically* illuminating. However, the argument I shall pursue is *not* that travel to and in Greece is essential to study early Greek philosophy, although my many years of visits have proved to be relevant and valuable for *my* research. My argument is, in part, that sensitivities concerning a panoply of cultural contexts have importance, vastly underestimated by scholars working in the field, and that the disciplinary indifference to visiting Greece is emblematic of this insensitivity.

Naturally, if the case can be made for the relevance of archaeology to the speculative thought of the earliest Greek philosophers by showing the usefulness of archaeological evidence at least to those originating episodes before the role of "philosopher" had been named let alone secured, then the value of context (and perhaps also visits to Greece) may take on an unexpected relevance. Also, let me make it clear that I am not arguing that such an approach makes a claim to supersede traditional approaches to research in the field, for it is patently clear that so much valuable research requires neither attention to cultural context nor travel to Greece. I am arguing only that the new approach advocated here will contribute positively to the exegetical goals of traditional studies and deserves greater exploration. I now turn to explain further why archaeology has been neglected by scholars of ancient philosophy.

"Archaeology is a mode of cultural production of the past."[7] It is a field of study embedded within historical and cultural contexts. Stated differently, the business of archaeology is usually identified with and grounded in "material culture," while the business of philosophy is customarily identified with the world of thought, the interrelation of ideas. That is, the business of philosophy is rather "nonmaterial culture." Hence, even on brief reflection, it quickly becomes apparent why the study of archaeology has been routinely ignored by scholars of ancient philosophy. Archaeology and philosophy investigate objects of a very different order. And if archaeology has been neglected, it is because its investigations are into a realm that, however interesting, is nevertheless irrelevant prima facie to what is "philosophically" meaningful. Thus, archaeology has been neglected as a possible resource for scholars of ancient philosophy because it has been perceived as a nonstarter.

On the contrary, the case I claim to have made is this: Archaeology—ancient artifacts and reports on them, and its mode as a cultural production of the past—has some relevance to philosophy, certainly with regard to an understanding of the historical origins of philosophy. This is especially true in accounting for the origins of the role of "philosopher" against a background of existing social roles. For we need to ask, again, *against what background of existing social roles in archaic Greece does the emergence of the "philosopher" represent a meaningful departure?*[8] The more deeply we investigate this issue, the more our discussion becomes a dispute about the nature of philosophy, on the one hand, and, on the other, the appropriateness of these debates in the specific ancient context from which it emerged.

In Part I of this book, I proceeded by accepting a position that I had outlined in *Anaximander and the Architects*: "Philosophy" is minimally demarcated by the adoption of rational accounts and the rejection of *mythopoesis*, evidence of the employment of rigorous proof as an acknowledged and required part of the process of securing a conclusion, and evidence of second-order questions in the testimony itself, that is, a self-reflective moment in which the thinker shows awareness of ambiguities in his or her own investigation.[9] With this platform, I then tried to show that when we review the surviving, tertiary testimonies about the Milesians, for example, we find many passages that have, historically speaking, presented difficulties of interpretation. By appeal to archaeological reports and evidence, however, I showed that a new light can be brought to bear on these textual and interpretative disputes—we get new and penetrating insights about the world in which these thinkers thought and, in some cases, the processes by which they thought, processes that mirrored the technologies of production. This is especially true of the Milesian *physiologoi*. By locating the historical objects and processes Anaximander believably witnessed, referred to in the surviving doxographical sources, we come to consider new hints about his processes of reasoning and speculation. The appeal to analogical and metaphorical reasoning to explain early philosophical thought has been conventional,[10] but no one before has tried to locate and present the archaeological evidence upon which some of the analogies and metaphors rest. When we see this archaeological evidence, we begin to place ourselves back into the lives that these thinkers experienced, and most especially to understand the variety of techniques that they and their communities, who would have been their audience, could have observed that produced their material culture and, quite literally, *grounded* their speculations. In these considerations, we have a promising approach for new insights about the early thinkers. Thus, we come to see the relevance of archaeology to the study of early Greek philosophy.

If this case has been made, even provisionally, a next step might be to see how the philosopher can hope to offer something in return to the archaeologist. What can the philosopher offer the archaeologist? The answer is that based upon the new insights gained about matters of textual interpretation, thanks to the archaeologists' reports and evidence, we will gain deeper insights into the belief systems of the early Greek philosophers. In turn, philosophers can now discuss with the archaeologists what new sorts of things, new pieces of evidence, need to be looked for. For example, a next step might be to catalogue all the references to *technai* that can be gathered from our doxographical sources, and from other cultural sources in archaic and classical Greece. Then, we can bring to our meetings with the archaeologists our lists of those machines and techniques that are mentioned, and the cultural matrices that funded, supervised, and developed these industries. These factors would be new to the discussion, relevant to it, and invite specialists in the field of ancient philosophy to explore archaeological resources and, thus, historical and cultural context.

EIGHT

What Is the Archaeologist's Theoretical Frame When Inferring Ideas from Artifacts?

A Short Historical Overview of Theoretical Archaeology

A

HOW IS ARCHAEOLOGY RELEVANT TO A PHILOSOPHER'S MENTALITY?

THE OBJECTIVE of this chapter[1] is to provide an overview of the development of archaeological theory with the principal intention of inviting readers to think again about the meaning of an "object"—whether a temple, a wheel, or a bellows—and to engage archaeological theory to get clearer about what kinds of constraints there ought to be on the narratives we tell by means of them. I am not in any way pretending here to give an exhaustive account of the trials and tribulations that have visited that discipline over theoretical quandaries.[2] But my purpose is, after giving a synoptic overview, to make clear through the rest of the book the metaphysical issues that arise from the methodological approach I adopted in Part I by using archaeological resources.

Let us review the *kinds* of evidence to be accounted for. In Part I, chapter two, we debated the meanings of the temple column and its construction. We focused on the symbolic meanings that might have been conveyed by vase paintings; then, we explored the parts of the column, the 3 × 1 proportions, the architectural techniques for modular thinking, and the possible role that column drums or bases contributed to symbolic thought. In chapter three, we distinguished architectural techniques of plan and elevation imaginations, and the symbolic meanings of 9 both from archaic poetic formulas in Homer and Hesiod and in proportions of Ionic temple column height, to shed light on Anaximander's cosmic numbers. In chapter four, we explored the various surviving images of the ancient bellows and its operations as we explored the cosmic mechanism by means of which the heavenly, fiery wheels radiate and the whole cosmos is itself kept alive by "breathing" through its blowholes. In chapter five, we investigated wheel making, axle structures, and wheeled vehicles to introduce in a new way how Anaximander's cosmos has hollow-rimmed wheels and turns along an axis of the world, an *axis mundi*. Finally, in chapter six, we reviewed what we know about ancient shadow-casting devices to think again about the lost seasonal sundial with which Anaximander is credited; the reconstruction of this solar device requires that we think through the steps and insights by means of which Anaximander imagined the most distant heavenly wheel and the cosmic patterns it displays in shadows. These, then, are the *kinds* of archaeological issues that need to be reviewed in their theoretical context. How do we make sense of the claim that the narrative presented in Part I that draws on archaeological objects in their historical and cultural context is in some significant ways "better than" other competing narratives? To set up the ground for an answer to that question, let me remind the reader of the point of view that has been adopted throughout.

The narrative of Part I displays the "Method of Discovery": Its narrative places the author at the temple building sites, standing next to an archaeologist, and next to both of us is "Anaximander," the Anaximander we know from the surviving doxographical reports and his belief system that we have inherited from the assessments of generations of scholars, modified by recent studies. When we explore temple column construction, we place ourselves at the ancient building sites quite literally, and then we begin reporting what we observe and are "reconstructing," acknowledging that many of the temples are now in ruins. Then, we refer the observations and proposed reconstructions to the context of other contemporaneous temples, often better preserved in parts, and then call upon the literary record when available to infuse our hypotheses about the stages by which the temples were built—the cosmos for Anaximander was also "built" naturally in stages. I am proposing this narrative to be a significant part of the background, almost completely unexplored in the scholarly literature, against which Anaximander's cosmic reflections—his abstract and speculative thoughts—acquire an illuminated meaning.

Looking at a prepared column-drum face, for example, the archaeologist is given the burden not only to explain what we are observing in the concentric lines around the face (*anathyrôsis*) and the hole in the center (*empolion*), but also the techniques by means of which this design and engineering were effected, the techniques that reveal not only why, but how, the drum face was so prepared.[3] Thus, scholars are now being invited to imagine that the *three* of us—philosopher, archaeologist, and "Anaximander"—are watching a workforce with a series of substantial tasks that must be completed in a specific order in order to affect the successful building of a monumental temple, a building celebrating a cosmic power. At some stage in the process of observation, I am suggesting, Anaximander had an epiphany: The temple of the cosmic powers appeared in some *analogy* with the cosmos itself. It might very well have dawned on him as the community leaders discussed and debated with the architects what symbolic designs were most appropriate to celebrate Apollo at the temple of Didyma, or Artemis at the temple in Ephesos, or Hera at the temple in Samos. In any event, the sequence and techniques of monumental architecture illuminated cosmic structure, and these sequences and architectural techniques held insights for Anaximander into cosmic processes. And, unlike Aristotle who reached the conclusion that there were two kinds of *phusis*, or nature, the one terrestrial and the other celestial, Anaximander observes techniques at the earthly building sites and then projects them onto the heavens because he envisages the earth and heavens as inherently interconnected. Whereas Aristotle came to regard the motion of the four "earthly" elements of earth, water, air, fire—terrestrial objects—to be rectilinear, straight-line motion to their natural places, he reached the conclusion that heavenly objects, because they moved in (apparently) perfect circles, while nothing on earth does so, were made of a different, fifth element that he called "aether." Thus, Aristotle inferred that celestial material was different from terrestrial—and not merely terrestrial material behaving differently at celestial heights—because of the distinct motion it displayed. Aristotle inferred the nature of matter from the motion exhibited. Anaximander's cosmos, in contradistinction, is physically interconnected; terrestrial laws and patterns are echoed by and in celestial motions. The view that heavenly and earthly physics are one *phusis*, subject to the same law-like patterns, is the presupposition for and basis of the microcosmic-macrocosmic argument that the structure of our experienced world mirrors the structure of the cosmos.[4] Thus, Anaximander's philosophical rationality allowed him to infer from the discovery of the natural patterns— the laws, revealed by the building of temples, making of wheels for transport, and operating the bellows at the forge—natural and law-like patterns consistent throughout the cosmos.

As the archaeologist offers an explanation, we imagine that Anaximander watched the *same* techniques. And even if we accept this first part of the account, we must ask further whether and on what grounds we have the justification to assert that Anaximander, observing the so-called same techniques, understood them as our narrative suggests. Since we already have before us in Part I the historical narrative by means of the archaeological details, we may set our exegetic sights through the course of Part II of this book on exploring the metaphysical commitments that are presupposed in and by this historical narrative.

Our narrative, then, relies on theoretical approaches in archaeology with sensitivity to issues in cognitive archaeology—an archaeological approach to "the study of past ways of thought as inferred from material remains."[5] But our narrative is different from studies in cognitive archaeology where almost all the literature focuses on prehistoric matters, and it is different from general archaeological studies since the archaeological narratives must yet be applied to the doxographical reports on Anaximander. In contradistinction, Anaximander lived in historical times, is referred to by later authors, and wrote a book that, at least in summary form, was the basis of doxographical reports. It is our objective here to try to infer Anaximander's ways of thought by means of the material remains that are purportedly referred to by him in those later reports. And the success of this approach is to be measured by the additional clarity brought to long-standing debates about Anaximander's philosophical rationality by appeal to archaeological artifacts and reports on those ancient techniques.

B

A SYNOPTIC OVERVIEW OF
ARCHAEOLOGICAL THEORY

Let me begin this section with a cautionary note. In attempting to supply a brief overview of theoretical archaeology, the risks of oversimplifying are great. The dangers lie in presenting different theoretical approaches as if the field of archaeology was monolithic—it is not. I mean that we should talk of *archaeologies* rather than archaeology, but, for the sake of my synopsis, I will speak of "archaeology" because I am trying to trace out a general thread. Let me try to be clearer about this point. In the 1960s, for example, there were continuing debates between culture history and processualism, between those who thought the business of archaeology was to produce historical frameworks and chronologies and those who believed it should do more, namely, investigate the *processes* of cultural and social changes throughout that historical framework. In the 1980s, new debates emerged between the processualists who were materialist, quantitative, hypothetico-deductive, and positivist in their approaches, and the postprocessualists who rejected positivism and its attendant methodologies. The processualists tried to fashion archaeology on scientific method, emphasizing the importance of testability or falsification of claims, in large measure because they believed that the past could be known in an "objective" way and that archaeology could explain it. The postprocessualists, deeply influenced by postmodernism, doubted that the past could be known in any "objective" way and united around a critique of processualism and its positivism; acknowledging the subjectivity of the archaeologist, they pursued, instead, intelligibility in their narratives. During these controversies, there were leading proponents who articulated positions that fueled the debates, and we will turn to consider them, but many archaeologists were not simply in one camp or the other. The point is that in presenting these theoretical debates, archaeologists adopted many variations and subtleties, and in the synoptic overview that follows, those variations and subtleties are glossed over. The situation is analogous among philosophers. While in the next chapter I will identify and subsequently criticize metaphysical realism, not all philosophers agree about how it should be defined. In following Kant's articulation of empirical realism and transcendental idealism, I try to make clear how I am understanding that position which Kant also criticizes. Quine's indeterminacy of translation, Davidson's radical interpretation, Putnam's internal or historical realism, and Searle's external realism all name positions about which philosophers share some agreement and disagreement. Accordingly, I cannot divide simply philosophers into those who accept the project of metaphysical realism and those who do not without risking gross oversimplification about the variations and subtleties of metaphysical positions that make up some of the central concerns of philosophy. Philosophy is no more monolithic than archaeology, and I want to acknowledge at the outset the perils of the forthcoming synoptic overview.

As previously noted, "Archaeology is a mode of cultural production of the past."[6] The data of archaeology are material culture. By collecting and inspecting ancient artifacts along with other historical indices, archaeologists attempt to tell us about ancient cultures, that is, archaeologists make inferences in order to produce their narratives. But can the archaeologists' discoveries allow us to gain insight into the *thoughts* of ancient people, indeed the earliest philosophers? Can archaeologists help us to gain insight into the mental processes and, in particular, the ratiocinations of people who lived long ago? And, if so, what are the limits or constraints on what can be inferred?[7] There has been, and continues to be, much debate within the community of archaeologists about what light archaeologists can shed in this regard, most especially in the arena of speculative and abstract thought.

In a curious way, this book is a kind of *inverse archaeology*. By this I mean that, thanks to a long history of classical scholarship, we know some things about Anaximander's belief system, and the Milesian *phusiologoi* in general. While most of the evidence is in the form of tertiary reports, there are good grounds for maintaining that some of the technological products that survive as the archaelogist's evidence can be traced back directly to Anaximander, pointed to by the use of certain technical terms that are plausibly his own. Starting with the outlines and details of Anaximander's belief system, can we infer the material stuff that inspired it, and to which Anaximander explicitly referred? In Part I, we began with the belief system and then appealed to the archaeologist's artifacts and reports to further clarify the belief system itself. This approach is the kind of *inverse archaeology* that I am suggesting; instead of beginning with the excavation of material artifacts and then trying to infer the ideologies that they suggest, we begin with a range of ideologies and then try to connect them to the material artifacts already excavated.

One recently emerging field of archaeology—cognitive archaeology—has attempted to address the problem of inferring "thought" from "material artifacts"—that is, the ideologies and mentalities of the peoples who produced the material artifacts that have been excavated. We shall now explore this development and its debate within the community of archaeologists.[8]

Cognitive archaeology developed in response to the so-called new archaeology of the 1960s and 1970s. After a while, when it no longer was "new," it came to be referred to as processual archaeology. New archaeologists criticized traditional (i.e., "old") archaeology—also referred to as *culture history* [9]—as purely descriptive and interested only in producing historical sequences of "cultures" constructed on the basis of artifacts generally found together. "Culture history" is both a goal and an approach (not a method) to achieving that goal.[10] It is probably easiest to think of "culture history" as shorthand for "cultural historical reconstruction." As such, and as a goal, culture history is basically event history: chronology (built through stratigraphy, radiocarbon dating, ceramic sequences, etc.) that delineates the origins and development (and decline, if relevant) of a site or a region or a culture. Culture history is the bones or framework or armature of archaeology; it is the who, what, when, and where of it all. It is inductive, descriptive, and normative: amassing facts and synthesizing them. The argument that led to processual archaeology back in the 1960s was that culture history is not *all* that archaeologists should be doing; the critics referred to old archaeology as "sterile" culture history; they objected that the archaeologist should be focusing also on the why and how—the processes of cultural and social changes throughout that historical framework. To do that, new archaeologists urged the use of a scientific, deductive, hypothesis-testing methodology to try to understand causal phenomena. Thus, new archaeologists, such as Lewis Binford and David Clarke, sought to incorporate the procedures of the natural sciences and anthropological perspectives into archaeology. Their goal was to develop an objective, scientific archaeology that would be able to explain the *processes* behind cultural changes[11] evident in material remains. Processual archaeology treats cultures holistically and attempts to produce general, testable theories based on deductive reasoning. It is positivist in outlook and concerned almost exclusively with material remains.

The materialist foundation of processual archaeology, with its emphasis on settlement patterns, subsistence strategies, ecological settings, and productivity, was a source of some archaeologists' dissatisfaction with new or processual archaeology, which led, in turn, to the development of cognitive archaeology. Binford dismissed attempts to explore ancient cognition and symbolic systems as "palaeopsychology." And yet, without taking the mental processes of individual members of societies into account, processual archaeologists are essentially left with a behaviorist theory that can only account for cultural change through appeals to environmental change.

Two approaches have arisen in response to this perceived neglect of the importance of human cognition in cultural systems. The first may be called cognitive-processual; it accepts the fundamental principles of processual archaeology (viz., scientific objectivity, hypothesis testing, etc.) for the most part, but attempts to apply its methodology to cognitive, as well as physical, aspects of culture. The second approach, dubbed postprocessual by its advocates, rejects the basic tenets of new archaeology. It accords just as much importance to cognitive systems and individual members of societies as it does to external stimuli and social norms in shaping cultures.

Colin Renfrew is perhaps the most influential of the cognitive-processual archaeologists. Renfrew emphasizes that thoughts from the past can never be recovered, but that ways of thinking may be inferred from material remains. Fundamental to this premise is Renfrew's notion of an individual's mental *mappa*, a cognitive map or worldview, that is based on experience with the external world and shapes interactions with it. While each *mappa* is idiosyncratic in some respects, Renfrew asserts that *mappae* must share some universal elements, otherwise meaningful interaction between individuals would be impossible. A key concept in Renfrew's theory is the idea that members of a single community (i.e., contemporaries in a given culture and locale) will have more common elements in their *mappae* than those that are universal.[12]

Mappae may be discovered by examining the way material objects reveal (or suggest) how individuals manipulate cognitive matter (symbolic systems) to produce given objects. Renfrew suggests six primary ways in which people manipulate symbols in their interactions with the external world.

1. People design things; that is, they manipulate cognitive matter intentionally to achieve a purpose.
2. People plan; they take time into account and sometimes produce physical diagrams in the process of implementing a design.
3. People use measurements; they use devices to quantify amounts of time, space, and mass and systems with related, comparable units.
4. People use symbols to help order social structures and personal interactions.
5. People use symbols as a means to relate to the supernatural.
6. People literally represent things symbolically; that is, they depict them physically.[13]

One of Renfrew's best known applications of the cognitive-processual method involves a series of weight-stones and pans found in the Indus Valley. The colored stones (apparently imported from a distant source) were worked into cubes and have masses in multiples of a constant (0.836 g) in a base-16 system. Renfrew drew the following conclusions from the stones:

(1) the people who produced them had a notion of equivalency similar to ours, (2) the apparent use of units entails the notion of measurement, (3) the use of multiples of a constant implies a notion of numbers in a system, (4) the association of pans with the stones implies a practical use for the stones using the measurement system posited in 2, (5) the idea of equivalency between the masses of different materials suggests a notion of relative value, and (6) the notion of value suggests a concept of rates of exchange.[14] James Bell has noted the ways in which these statements adhere to the cognitive-processual method. The statements are *about* thinking, making no claims to recover actual thoughts. All the statements either arise directly from the data from the artifacts or follow logically from others that are tied directly to such data. The statements are thus testable, inasmuch as other finds of weights that did not fit into the hypothesized system would prove them false. Finally, all the statements are logically consistent and not dependent on subjective points of view.[15]

Joyce Marcus and Kent Flannery are new world archaeologists who have used cognitive-processual techniques to explore the importance of cosmology in the development and layout of Zapotec population centers. Marcus and Flannery have derived information about Zapotec religion from surviving precolonial maps and codices, as well as ethnographic documents prepared by Spanish colonial authorities in the sixteenth and seventeenth centuries. These documents describe Zapotec religious practices and iconography at the time of Spanish colonization and suggest what material remains archaeologists should look for as indicators of religious activities. Using such ethnographic data (which they call the direct historic approach), Marcus and Flannery were able to plot the development of a complex Zapotec state with the development of formalized Zapotec religion distinct from that of other Mesoamerican ethnic groups. They thus utilize ethnographic texts in combination with contextual analysis of material evidence of cult practices and patterns of public architecture to explain the rise and development of the Zapotec state.[16]

Next, let us turn to consider another study in cognitive archaeology, this time from a postprocessualist. As the reader will see, on the one hand, the basic archaeological techniques are not different from those of the processualists,[17] but, on the other, the theoretical frame and expectations of interpretation do indeed differ.

In a series of studies, J. David Lewis-Williams provided an in-depth view of the San people, bushmen from southern Africa, who left a record of rock art. His work also contributed to rethinking the famous Upper Paleolithic Franco-Cantabrian cave art, like that in Lascaux. The way he tells the story also helps to show the postprocessualist approach he adopts.[18] His narrative shows how archaeologists appeal both to "reception history" and "culture memory."[19] Reception history or reception theory is not a unified study but rather represents diverse discussions that ask us to pay attention to the history of the meanings that have been imputed to archaeological monuments and artifacts, as well as historical events.[20] Reception history traces the different ways in which various retrospective interpreters such as participants, observers, and historians have attempted to make sense of monuments, artifacts, and events, both as they unfolded and over extended subsequent periods of time, to make those events meaningful for the present in which they lived and live.[21] A second issue that bears on historical narratives because it helps us understand what recommends one narrative over another is what Jan Assmann referred to as "cultural memory"—the outer dimension of human memory.[22] He makes a twofold division distinguishing between "memory culture" and "reference to the past." *Memory culture* is a way of preserving cultural continuity from one generation to the next. *Reference to the past* is a way of considering how a society secures its collective identity by creating a shared past.[23] To refer to cultural memory, then, is to consider the underlying agendas, processes, and social predispositions by which the selection and dissemination of "approved" historical narratives transpire.[24] By this I mean also that appeals to cultural memory are not intended to establish what counts as an "accurate" statement in an historical account but rather to consider whatever has become persuasive to a particular group of individuals, the assumptions and prejudices that are shared by both the writer of an historical narrative and that writer's audience. An appeal to cultural memory seeks to reveal the predispositions of a group, whether or not the particular narrative is compelling or even rational. Lewis-Williams showed us how the reception history of Franco-Cantabrian cave painting reflected the changing faces of cultural memory.

During the first half of the twentieth century, as Lewis-Williams presents his narrative, the first view of the Franco-Cantabrian cave paintings was that they represented "art for art's sake." The supposition was that human beings felt an inner need to express themselves. Not long after, led by Abbè Henri Breuil (1940s and early 1950s), who appealed to ethnographic analogies, the new hypothesis that gained currency was that the paintings were motivated by hunting and sympathetic magic. Aided only by stone weapons, bringing down a large bison was no easy achievement; if the hunters could imagine in advance the hunt—the cooperation and coordination of efforts—it was believed that they could improve their chances for success. Thus, Breuil's thesis was that the cave paintings were made to sustain the material basis for life. Subsequently, a new approach was heralded by the introduction of structuralism in the work of Annette Laming-Emperaire (1962) and Andre Leroi-Gourhan (1968); their approaches postulated a "mythogram" or conceptual template that was emblematic throughout Upper Paleolithic times. Sharing an approach championed by Claude Levi-Strauss, they argued that a supposed universal binary system of thinking directed the patterns of human thought. The structuralist theory was that the mythogram thought itself through the minds of people into art. What all three of these approaches share is the view that suprahuman forces—not individual agency—were responsible for the cave paintings and rock art. First, it was nature's own force that drove human beings inexorably to artistic expression; then it was the force of nature expressed as human striving for the sustenance of material life that motivated the artistic expressions; and then, structurally, it was the biologically programmed patterns of human cognition that were responsible for the paintings. Lewis-Williams' narrative, then, traces the changing faces in the reception history of the Franco-Cantabrian cave paintings; the different views reflect aspects of culture memory, as well as the changing social and political agendas of those who controlled the dissemination of interpretations.

In the 1980s, Renfrew's work in cognitive archaeology, and also that of Hodder (1992) and Shanks and Tilley (1987), took a renewed look at the role of human agency. They were influenced by Pierre Bourdieu's theory of practice and Anthony Giddens' related structuration theory (1984). Both the approaches of Bourdieu and Giddens rejected a suprahuman supposition and tried to reconcile the complex relation of individuals and society in terms of mutual interaction; in that scheme of things, some individuals were supposed to understand society well enough to manipulate it, while at the same time being constrained by overarching social rules and resources. But Lewis-Williams' new hypothesis in understanding Upper Paleolithic cave paintings by means of San rock painting resulted from his approach to human agency and the complex interplay he detected between constraint and enablement in a society. To new ethnographic analogical reasoning, provided by Megan Biesele in her studies of the Ju/'hoansi people of the northern Kalahari Desert, Lewis-Williams credits his own breakthrough. Despite the fact that the Kalahari people were distant in geographical location and language groups, Lewis-Williams concluded that major beliefs and rituals were virtually identical to the nineteenth-century San peoples who were the last to leave rock art. Based on this ethnographical analogy, he reasoned that San rock art was in large measure associated with shamanic rituals, symbols, notions of supernatural powers, and spiritual experiences. The images on the rocks were expressions of trance states entered by the shamans. His thesis was that groups of shamans banded together to extend their influence and exploited their religious functions in order to acquire political power. His argument rested on what he regarded to be the multiple and empirically verifiable fit between the ethnographic analogy of the Kalahari Desert people, who still maintained such practices and features of San rock art. And this in turn allowed him and others to reopen the debates about the meaning of the Franco-Cantabrian cave paintings. To understand this new connection and insight, let us briefly consider ethnography and ethnographic analogical thinking.

Ethnography is the study of a group of living people in the present; by means of it, archaeologists try to gain clues about how people lived in the past. The ethnographer studies how people live, how they interact, what they believe, how they behave, what kinds of objects they use, and what they do. Ethnographers—sometimes known as "cultural anthropologists" or "behavioral anthropologists" or "social anthropologists"—tend to focus on one group at one place in time and generally spend a great deal of time living with and interacting with the group of people they are studying.[25]

The problem of understanding the ancient people who lived long ago is obvious. We cannot live with them. By appeal to ethnography and ethnographical analogy, archaeologists try to understand materials from the past. But there are also many problems and issues associated with this practice; this is especially true because ethnography and ethnographical analogies cannot constitute a proof about past practices. Let us consider some cautionary considerations.

First, there are many phases in the interpretation of artifacts—that is, taking an artifact as it exists in the ground and then understanding how it was used by living people in the past. One of the earliest phases in this process occurs when we classify the artifact as one thing or another, or as coming from one period of time or another—when we decide, for example, on a vase painting, whether a column is really a column or something else, whether a cup is a cup or a vase or a trophy, whether a clay figure is a toy or a religious object or a grave good, or if a horseshoe was used as a game, a piece of farming equipment, or a good luck charm. This early stage of work is an interpretive process; "classifying" or "sorting" is an interpretative act.

When archaeologists use information about present-day cultures to gain insights into possible ways that ancient peoples used or classified objects, they are engaging in ethnographical analogical thinking. The perils of this approach are many, for there can be no guarantees that the ways people today use objects and classify them were practiced by people long ago. However, by looking at examples from the present, archaeologists can ask certain questions about an artifact from the past. When an object is found in an excavation, we begin to wonder what it is. When we suspect it is one thing, we begin to wonder what might be found nearby. We will likely continue to wonder what would be found over it, under it, and so on. We might reflect on who used it and what he or she may have done with it.

Thus, while ethnographical analogy often proves useful, we cannot mistake it for proof. We understand, of course, that over time and space people change the ways they use things. Just because an object is used in a certain way in the present by a specific group of people, there can be no assurance that it was used the same way in the past by a different group of people, even if descended from them. The archaeologist is not in a position to say that just because the so-and-so people use an object in a certain way then *eo ipso* another group of people in the past did the same thing. Consequently, ethnographic analogy, despite its usefulness in assisting us to ask good questions, does not offer a sufficient condition to "prove" how something was or was not used.

Archaeologists use ethnographical analogies to assist in formulating questions for their inquiries and to test their hypotheses in much the same way that they use historical documents, oral histories, and other sources. Ethnographic analogies bring archaeologists into the *hermeneutic circle*. In the process of asking questions about an object, archaeologists understand more and more about the *contexts* of the *artifacts*, the *site*, and more, and with these additional insights we return to ask more questions about the object, and with the new insights ask more about the contexts, sites, and so on—the interpretive process never really ends. Archaeologists come up with their "findings," but if one adopts a postprocessual approach, these findings, in principle, can never be more than "provisional." This is not to say that there is no gain in understanding, but that the project always remains revisable, subject to debate, new evidence, and further reflection.

Thus, archaeologists move back and forth between their data—*artifacts* and artifact *contexts*—and their theories. When they go back and forth this way, they sometimes realize that their original ideas about what artifacts "were" in the past need to be changed. This is why the original sorting system, or classification system, is looked at as only a first step in understanding the artifacts and artifact contexts that the archaeologist finds in the ground; of course, sometimes the questions themselves also change over time.

In Lewis-Williams' study of San rock art, we can see the postprocessualist archaeologist at work. His interpretative archaeological approach showed how theories about Franco-Cantabrian cave paintings changed from art for art's sake, to sympathetic magic for the hunter to sustain the material basis for life, to structural cognitive patterning, and then to artistic expressions borne of shamanism and trance states to arrogate civic authority. Different theoretical approaches were inspired by changing cultural inclinations during different times by archaeologists with different agendas. And ethnological analogical thinking, despite its inherent problems, opened the doorway to the latest round of theoretical interpretations promulgated by Lewis-Williams. In a curious and inverted way, our narrative about Anaximander bears a resemblance to ethnographic analogical thinking: The architect-excavators are the archaeologists who represent the present "living people"; they specialize in architecture and its related technologies and provide narratives of the ancient building techniques. Following the ethnological analogy, the approach here supposes that long ago Anaximander was watching the architects and their workers engaging in the processes they have articulated. How Anaximander interpreted and projected metaphorically onto the cosmos what he discovered at the building site, however, is beyond the scope of the archaeologist per se. We shall investigate the philosophical search for meaning—interpretation and imaginative projection—in chapters nine and ten. But first, we shall examine in more detail the postprocessualist or interpretative archaeological approach with which our study shares the closest kinship.

C

POST-PROCESSUAL OR
INTERPRETIVE ARCHAEOLOGY

New archaeology, a primarily American school of thought, itself came under criticism in the late 1970s and early 1980s from theorists associated mainly with Cambridge University in the United Kingdom. Influenced strongly by European poststructuralism (especially the works of Derrida, Foucault, and Bourdieu), scholars such as Ian Hodder, Michael Shanks, and Christopher Tilley questioned the very notion of objectivity in research. Moreover, they pointed to an inherent problem in the empiricism of processual archaeology: Direct observation of the past is impossible, and no amount of quantitative analysis of artifacts can make up for that fact.[26]

Central tenets of postprocessual archaeology include the following:

1. The subjectivity of the archaeologist unavoidably plays a central role in the interpretation of the past.
2. Archaeology takes place in the present, constructing past histories based on interpretation of artifacts in the present.
3. Social practices and the objects they produce and utilize are meaningfully constructed.
4. There is never a definitive explanation for any given archaeological data.
5. Archaeology should seek to explain and understand meanings rather than develop theories of causality.
6. Multiple explanations of the same data serving various purposes from different subjective viewpoints are necessary.
7. Archaeological explanations are critical constructs that resemble performance in some ways.[27] Interpretive archaeology emphasizes the importance of context to the interpretation data and eschews cross-cultural generalization.

Hodder suggests that archaeological remains may be read like texts by interpreters who must always be aware of their own subjectivity. This notion of physical remains as text depends on the idea that humans produce meaningful symbols in some universal fashion. Meaning may be found in structured systems of functionally interrelated material remains. These material systems are seen to be the instantiations of public and social concepts that are reproduced in the practices of everyday life.[28]

The most important part of interpreting the past for Hodder is the recognition of systematic similarities and differences within a cultural context. Archaeologically relevant similarities and differences are manifest most often in terms of time, space, depositional unity, and physical typologies. Thus, archaeologists look at contemporary objects to hypothesize systematic relationships and noncontemporary objects—ethnographic analogies—to suggest change over time. Objects found together or within an area determined to be a unity in some way are likely to provide data for meaningful similarities/differences. Likewise, objects of similar shape and/or function may be meaningfully compared. None of these criteria, of course, is necessarily meaningful. The more meaningful similarities and differences an hypothesis is able to reveal across diverse aspects of a given set of archaeological data, the stronger that hypothesis is held to be.

Hodder utilizes Bourdieu's notion of cultural habitus to account for changes in cultural systems. Habitus is defined as socially constructed modes of perception and unconscious linguistic and cultural competences that allow individuals to cope with unforeseen situations.[29] Habitus is described as an intermediate link in recursive chains connecting social structures and individual practices. Individuals are affected by social norms, but do not act mechanically according to them. Rather, individuals factor in personal experience/opinion in weighing the potential costs and benefits of acting according to norms. Norms are thus constantly open to change brought about by individual actors in a system.

Processual archaeologists have objected that interpretive archaeology leads to undifferentiated relativism. Shanks and Hodder respond, following Bhaskar, by dividing relativism into two categories: judgmental and epistemic. Judgmental relativism holds that all theories are subjective and thus equally valid, whereas epistemic relativism acknowledges the subjectivity of the theorist, but, nevertheless, holds that theories are valuable only to the degree to which they correspond to the external data. Hodder and Shanks reject "judgmental relativism" and instead argue for "epistemic relativism" that allows for a meaningful distinction between "better" and "worse" accounts by appealing to the interconnected web of statements, evidence, and interpretations in which their theories cohere. One approach taken by interpretative archaeologists, then, is to advocate epistemic relativism. They can claim, therefore, to exclude fringe theories from serious debate, while arguing for the validity of diverse points of view.[30]

An example of Hodder's analysis of archaeological remains relates northern European stone grave monuments of the Neolithic period to houses of the same era. Hodder first describes detailed similarities between the tombs and houses in terms of construction techniques, spatial orientation, decoration, and internal divisions inter alia. He then cites a scholarly consensus that the Neolithic era of interest was characterized by small-scale, dispersed, acephalous lineage-groups.[31] Hodder next turns to ethnoarchaeological studies of such groups in east Africa to suggest that Neolithic Europeans were probably very concerned with increasing labor power by controlling women in the domestic sphere and female reproduction. The ritual associations of tombs and their apparent similarities to houses (with emphasis on the female quarters) are seen as attempts to give legitimacy to resource claims by appealing to ancestral authority. Hodder's exegesis thus not only describes the development of the physical form and function of the megalithic tombs, but also places them in a particular context and explains their symbolic functions.[32]

D

SOME CONCLUSIONS ABOUT ARCHAEOLOGICAL INTERPRETATION

The differences between cognitive-processual (e.g., Renfrew) and postprocessual archaeologists (e.g., Hodder and Shanks, and Lewis-Williams) appear to be mostly epistemological and ideological.[33] The former, while acknowledging limits to the scientific method and the inherence of subjectivity in the scholar, continue to insist on testable theories as a way of describing the past in a more or less "objective" manner. Here the term "objective" suggests that there was, in fact, a way that things happened and that an account could be mapped onto those "real" events, corresponding to those events. The latter refuse to subscribe to the notion that the past can be known in this sort of objective way or that any explanation of it could be definitive, that is, a last and final word no longer subject to revision. Still, postprocessual archaeologists try to preserve a meaning of "objectivity" by inviting us to see it as an account that resists specific trials of challenges without there being a single definitive account that could supersede all others. An "objective" account for postprocessualists like Shanks and Hodder acknowledges that "reality" is plural, that artifacts are multiplicities, and thus the meaning of a thing depends on what work is done on and with it.[34] The resultant explanations of archaeological data, however, do not seem all that different. This similarity may stem from apparent contradictions within each method in which the adherents necessarily use methods that they criticize in the other school of thought. Cognitive-processualists like Renfrew admit that theories about the past, and especially about past cognition, are ultimately impossible to demonstrate conclusively. They also admit that scholars necessarily bring subjective viewpoints to archaeological problems. It is unclear, then, how they can claim any privilege for their theories as more objective or somehow "truer" to the past. On the other hand, postprocessualists like Shanks and Hodder claim to avoid the devastating charge of acknowledging absolute relativism—which undermines all meaning—by insisting that theories *fit* the archaeological data as closely as possible, without ever providing criteria for "closeness." Thus, they also make an implicit claim to achieve a kind of "objectivity" that looks a lot like a coherence theory of truth. Let us consider this matter in more detail.

Hodder and Shanks argue that objectivity is something constructed, and explain further that it is crafted: "Objectivity is heterogeneous networking—tying as many things together as possible."[35] What they argue for is a kind of coherence theory of truth. A statement about the archaeological past proves to be "objective" on their account, not because of logical coherence or correspondence with something external, but rather because it "holds together" when interrogated. By this they seem to mean that statements about the archaeological past are tied into a web of interconnected claims about the excavation sites, the earlier states of knowledge about the finds there and elsewhere, prevailing interpretations, and ongoing debates. When such statements are interrogated, they either hold together, amidst the challenges, or they fall. But when they fall, they threaten to collapse other claims with which they are interconnected and on which they themselves rely. Thus, *objectivity* turns out to be *what is held together*. Stated differently, objectivity is *what coheres*. The archaeological past will not excavate itself, of course, but must be worked for; the archaeologist excavates, interprets, and provides statements that can be tested and challenged within the network of claims. In these senses, objectivity is something constructed.

In concluding this chapter, let us be clear about the interpretative archaeologists' position on the "reality" of the past, and hence the metaphysical commitments that are presupposed by their narratives. For Hodder and Shanks, for example, is there any sense of a *real* past on their account? The problem is this: In what sense(s) can the past be *real*? What can be the meaning of "real" if reality is socially constructed? Their position is that such questions are both wrongheaded and misleading. Interpretative archaeologists hold that reality is plural, and that artifacts are multiplicities; the meaning of "real" will depend on the context in which it is used, with regard to the work done on and with the artifacts. The mistake in such questions is to suppose that there can be a "real" past in the *singular* sense, some ultimate frame of things onto which our narratives can be mapped definitively. The reality for which they argue, then, is not an absolute any more than the objectivity they seek. "Reality" and "objectivity" are not absolute or abstract qualities.[36] The account that best resists the challenges of other competing theories, and the challenges of evidence, is the one that enjoys the designation of "objectively better." But when and if new evidence and new challenges arise, the archaeologist's narrative survives or falls, and with it the series of interconnected claims. When a narrative is challenged, then, Shanks and Hodder recommend the excavation of another site, the mobilization of another army of people, heterogeneous mixtures of peoples and things, to see if the new results challenge or lend support to earlier claims, to see if they stand up to a diverse challenge from a range of multiple perspectives. This is the best we can hope for when we produce our narratives by appeal to archaeological resources.

NINE

The Interpretative Meaning of an Object

Grounding Historical Narratives in Lived Experience

John Kaag, co-author

―――――――

NOW THAT WE have surveyed theoretical approaches to archaeology with an eye to our narrative on Anaximander that makes use of archaeological resources, we turn to explore the philosophical underpinnings of this project. As will become clearer still, our study shares a philosophical kinship with proponents of postprocessual or interpretative archaeology. We have addressed two underlying issues throughout Part I: The first involves interpreting the meaning of an object, and the second is the process of mentally projecting from our material context to another domain of imagined, abstract thought. Anaximander watched activities at the temple building sites and interpreted the meanings of objects in a multiple fashion. He saw a column drum and came to view it as representing the shape of the earth. He saw vehicles of transport and came to see the wheels suggest the sun, moon, stars, and the cosmic axle. He saw the operation of the bellows at the forge and came to see it as supplying the mechanism for the fiery radiation and cosmic respiration. Stated generally, then, our two underlying issues are: (1) how do we determine the meaning of an object, archaeological or otherwise? and (2) how is it that we connect our material-embodied experience to the domain of imagined thought?

The exploration of these questions takes us to hermeneutic and pragmatic interpretations.[1] And when these approaches have been explicated, we find ourselves staring at the metaphysical foundation of our philosophical project. The examination of the metaphysical underpinnings of the project takes us to arguments such as Quine's indeterminacy of translation, Davidson's radical interpretation, and Putnam's internal or historical realism. Finally, we address more directly Searle's arguments for External Realism. Searle's arguments join

company with other approaches that have sought to defend broadly the project of metaphysical realism. By the end of the chapter, it will be clear that the
project of metaphysical realism is as hopeless as it is wrongheaded. This stance
points the way to a new set of arguments for what I will call "experiential realism," delineated in the last chapter of the book.

A

THE IMAGINATIVE MEANING OF AN ARTIFACT

The analysis of the fragments of Anaximander detailed in Part I stands as
an attempt to mediate not only between the two seemingly distinct fields
of archaeology and philosophy, but also between disparate philosophical
schools. It aims to appeal to the analytic rigor of philosophers in the Anglo-
American tradition, but insists that this rigor must be tempered, tested, and
indeed enlightened, by a historical and sociological approach to truth
claims.[2] This latter approach characterized the continental and pragmatic
thought of the nineteenth and twentieth centuries. The work of Hans-
Georg Gadamer reflects a sustained interest in the historical and cultural
context of thought, a willingness to use this context in order to interpret
and reinterpret thought, and a suspicion that such reinterpretation creates
an opening for the meaning of a given thought to emerge. Along these
lines, American pragmatists such as John Dewey and Charles Sanders
Peirce, in marked contrast to Barnes' positivistic stance, insist that artifacts
are never *just* artifacts, that the objects of the world cannot be understood
by way of a single objective meaning, but only through a process of imaginative interpretation.[3]

   This interpretative and imaginative method is under way when cognitive archaeologists conduct their research and unearth complex meaning
from excavated artefacts. This imaginative interpretation, however, does
not fall prey to thoroughgoing relativism but instead provides a way of producing multiple possible interpretations or hypotheses in regard to the
meaning of an object; these hypotheses stand to be tested by the wider
material circumstances of future archaeological studies. As Dewey notes,
not all hypotheses reflect equal merit. In this section, the argument being
defended is that the processes by which archaeologists infer ideas from artifacts should be understood in the philosophical context of the genealogical
and hermeneutic approach of the continental school and the method of
inquiry set forth by American pragmatism. While this method may find its
philosophical roots in the continental and pragmatic schools, it is worthwhile to demonstrate the way in which interpretation has emerged in the
writing of analytic philosophers such as Willard Quine, Donald Davidson,
and Hilary Putnam. Their respective works on the "indeterminacy of trans-

lation," "radical interpretation," and "internal or historical realism" echo the sensibilities reflected by interpretive archaeology and begin to speak to the culturally contextual basis of thought.

B
HERMENEUTIC AND PRAGMATIC INTERPRETATIONS
B.1

DIGGING FOR MEANING: HERMENEUTIC PLAY,
INTERPRETATION, AND ARCHAEOLOGY
"WHAT *IS* THIS *THING*?"

To understand the character of hermeneutic play and interpretation, it is appropriate to examine the question and situation from which both arise. This question, one that interrogates the ontological status of a "thing" in the world, serves as the common refrain of most archaeological projects. This "thing"—as yet unknown—protrudes from the ground and invites an archaeologist, after some playful thinking, to take a certain educated guess as to what it might *be*, for example:

> This thing is stone and cylindrical in shape. This thing is actually quite large. This thing is six feet in diameter and two feet tall. It has a type of fluting around the edge. On each face, this thing has a smooth band running around its circumference, and another concentric circle etched closer to the center. The thing has a hole in the middle of its circular face.

These observations slowly fall into place for the seasoned archaeologist, who also relies on antecedent experiences of such objects found in other excavations. The archaeologist is drawn to a tentative hypothesis that emerges from the thoughtful play with the object. This is not just any old thing—this thing is a *column drum*. This is a move from the discovery of evidence—as a matter of fact—and an interpretation of this evidence that places this "fact" in a social, political, and cultural matrix. Such a hypothesis generates subsequent thoughts in regard to the meaning and architectural function of the column drum in the wider context of the excavation site. This brief vignette provides a snapshot of the hermeneutic play and the process of interpretation that leads one to recognize what an object *is*, and to grasp the meaning of a thing. In his *Truth and Method*, Hans-Georg Gadamer describes this process in detail, but also suggests the necessity of interpretation in exposing the truth of an object. For Gadamer, as for the cognitive and interpretive archaeologists, "things themselves" always point beyond themselves to a constellation of complex meanings.

In the first part of *Truth and Method*, entitled, "The Question of Truth as It Emerges in the Experience of Art," Gadamer explores the way in which a particular interaction with an object of art has the ability to expose the meaning and truth of that object. Gadamer wishes to show that this truth is neither objective in the sense of disclosing one true meaning of a particular object nor is it merely subjective and idiosyncratic. This interpretative approach to the truth of aesthetic objects might be fitting in the investigation of objects tout court and especially in reference to objects of ancient architecture. This move seems to correspond to Gadamer's own sensibilities. As other scholars have noted, Gadamer begins by applying his hermeneutic method to art, but quickly turns his attention to texts, to history, and to anything that is "handed down to us" through living tradition.[4] Archaeologists, such as Renfrew, Lewis-Williams, and Hodder, seem to extend and enliven this living tradition. Despite Barnes' positivistic protestations, archaeologists are always in the process of a type of hermeneutic play, a fact reflected in both the diverse interpretations of archaeological finds and the continued investigations of uncovered sites.

Gadamer suggests that the work of art engages us and draws us into an ongoing and interactive play. We are drawn in not as disembodied subjects, ready to pass unbiased judgment on a given object, but rather beckoned in as situated individuals whose prejudices provide the limiting, but also the enabling, conditions of playful inquiry.[5] The games we play with a work of art or object of history cannot simply be anticipated or reproduced in any deterministic fashion; instead, we give our attention to the object—or more accurately, to the play itself. In so doing, we allow the object to show itself, *speak for itself*, in a new and interesting way. As archaeologist and philosopher Allison Wylie notes in *Thinking Through Things*, archaeologists have the difficult task of interpreting and translating the language of objects; only through this task are investigators drawn inductively to a hypothesis that emerges from the particular play of thought.[6] Expressing a comparable sentiment, Richard Bernstein notes in his description of Gadamer and interpretation: "We should always aim at a correct understanding of what the 'things themselves' say. But what the things themselves say will be different in light of our changing horizons and the different questions that we learn to ask."[7] This comment seems to echo Hodder's insistence that archaeological evidence must be given unique readings and can give rise to disparate interpretations.

The character and being of an ancient architectural object shows itself through its interpretative conversation with a particular observer. In this case, what an object *is* can only be grasped by inferring what an object *means*, or, in the case of the column drum, what an object meant. For Gadamer, understanding an object and thoughtfully interpreting the relations that constitute the meaning of the object are one and the same. It is interesting to note that here Gadamer seems to echo the Parmenidean fragment that is enlivened in both the phenomenological and the pragmatic traditions: "Thinking and being are (of) the same."[8] The relations that are understood constitute what an object is and also what an object comes to mean. Cognitive archaeologists such as Renfrew and Lewis-Williams seek to unearth the relations, both cultural and psychological, that might have constituted the architectural remains. By the same token, Gadamer comments on the way in which these complex relations constitute the work of art, writing, "A work of art belongs so closely to that to which it is related that it enriches its being as if through a new event of being."[9]

Along these lines, a work of art, and, by extension, a work of architecture, is not to be thought of as a self-contained and self-enclosed object. It is not a thing-in-itself (something *an sich*) that stands against a spectator and which a spectator is to first purify himself in order to understand. Instead there is a dynamic conversation and interaction between an object and the archaeological audience that brings a set of meaningful relations into the open. Gadamer is well aware of the uncanny nature of this conversation; he is, after all, suggesting that examiners engage in dialogue with texts, with paintings, and with large stone objects. He is the first to admit that these objects "are permanently fixed expressions of life which have to be understood, and that means that one partner in a hermeneutical conversation, the (object) is expressed only through the other partner, the interpreter."[10]

This being said, it is not the case that the interpreter is free to express whatever he chooses in regard to the meaning of an object. In other words, Gadamer does not succumb to a type of rabid relativism, but rather insists that interpretations are grounded in a particular, real situation. The interpreter must be mindful of the object. His interpretation must be open to falsification through ongoing archaeological and, in the case of Greek artifacts, doxographical evidence. On this count, the cognitive-postprocessual interpretations of Lewis-Williams or even the cognitive-processual accounts of Renfrew and others should not be viewed as flights of fancy, but as earnest hypotheses to be tested and situated in the wider context study. Obviously, the more sweeping a hypothesis, the more likely it is that it will need to be revised— and this is why the archaeologists claim to need middle range theory[11]—but this fact should not discourage one from educated guesswork. Such a process of revision should be regarded—to fall back on the language of Gadamer once more—as the hermeneutic movement in which meaning continually emerges.

In the next section, we will explore the basis on which one might construct and judge particular interpretations. Interpretations can be judged pragmatically—that is, they can be judged on the basis of their consequences.

B.2

PRAGMATIC INTERPRETATIONS:
FROM MATERIAL CONTEXT TO IMAGINING THOUGHT

Gadamer is not alone in his belief that the constitutive relations of a thing— whether a work of art or an architectural feature—can be revealed only through a thoughtful interpretation of, and what he refers to as a conversation with, that object. For American pragmatists such as John Dewey, this conversation with objects highlights "the material and the ideal in their reciprocal interrelationships" and directly suggests that this reciprocity must be underscored in grasping the meaning of an object of history and the writings of particular thinkers.[12] Dewey's description of a natural scientist in *Experience and Nature* reflects this suspicion and may enlighten the general method by which archaeologists explain their findings. Dewey writes:

> A geologist living in 1928 tells us about events that happened not only before he was born, but millions of years before any human being came into existence on this earth. He does so by starting from the things that are now the material of experience . . . he translates, that is, the observed coexistences into non-observed, inferred sequences.[13]

In the case of postprocessual archaeology, Hodder and others "translate" from artifacts to the *possible* psychological and cultural conditions that arose in tandem with them.

This process of translation is not predetermined by a set of rules, but is motivated by the imaginative inference of a given investigator. This fact seems to be born out in the variety of scientific hypotheses. The variety of inferences, however, should not be confused with a type of unordered randomness. This point resonates with Hodder's response to objections made by processual archaeologists that postprocessual archaeology leads to radical relativism. Dewey, facing similar criticism, gives Hodder's defense philosophic teeth by claiming that investigators move from the "material of experience" to "non-observed" inferences by way of a particular process of embodied inquiry that can be judged in accord with its outcomes.

According to Dewey, "Inquiry is the controlled or directed transformation of an indeterminate situation into one that is so determinate in its constituent distinctions and relations as to convert the elements of the original situation into a unified whole . . . inquiry is concerned with objective transformations of objective subject matter."[14] Truth is to be defined in terms of inquiry and, correlatively, in terms of the unity that is the outcome of successful inquiry. This approach to "truth" is as unsatisfying for many analytic philosophers as it is for processual archaeologists; both groups claim that such an approach leads to either a vicious logical circularity or a type of unproductive relativism.[15] Dewey, however, is willing to flirt with these undesirable outcomes for the sake of describing a theory of inquiry that seems to account for the quality of particular indeterminate situations and for the satisfying consequences of arriving at the unified whole that characterizes particular forms of inquiry.

Cognitive-processual and postprocessual archaeologists embark on the movement of inquiry when an experienced situation is framed as a unique problem—a unique question, to be answered. To recall our earlier example, this "cylindrical stone" poses a specific question: "What am I?" For some investigators this situation would be settled and unified in the conclusion that this thing is a column drum, but the postprocessual archaeologist might carry this inquiry a step further. These scholars insist that the existence of a column drum poses a subsequent question: "What do I *mean?*" According to Dewey, ideas in reference to this problematic situation "occur at first simply as suggestions; suggestions just spring up, flash upon us, occur to us."[16] The way in which these tentative hypotheses and suggestions "spring up" is described in the early pragmatism of Charles Sanders Peirce, in his concept of "abduction."[17] Dewey fleshes out these abductive processes, explaining that they can be best understood as the movement of human imagination. A brief description of *imagination*, then, seems warranted and helps make sense of the interpretative dimension of postprocessualist archaeologists.

In 1934, Dewey published *Art as Experience* in which he examines experience as the "result of an interaction between a live creature and some aspect of the world in which he lives."[18] In this work, he addresses the imaginative process that often characterizes this interaction and suggests that the imagination may play a vital role in interpreting the material of experience. He opens this discussion by commenting that the imagination "is a way of seeing and feeling things as they compose an integral whole. It is the large and general blending of interests at the point where the mind comes in contact with the world. When old and familiar things are made new in experience, there is imagination."[19] Here Dewey expresses at least two important points in regard to the current examination of postprocessual and cognitive archaeology. The first concerns the holism reflected in the hypotheses of Hodder and others. Instead of viewing an artifact as a single moment of "material experience," Hodder and his colleagues acknowledge the way in which any object of history is situated in a constellation of related objects. The column drum might be discovered next to other architectural elements such as a piece of entablature or a capital. It might be found at the top of a hill or, perhaps more interestingly, in an open valley. By situating the object in a wider geographical and cultural setting, scholars begin to form educated guesses about the various relations that constitute the work of art. This interpretation aims to create a coherent and unified whole by imagining the contextual factors that gave rise to the meaning of the object. Wylie notes that this imaginative process takes its cues from the process of abduction described by Peirce, often considered the forefather of American pragmatism. Abduction, like the event of imagination, strives to revive older modes of thought and meaning in novel conditions.[20]

Dewey's comment that the imagination has the unique ability to make "old and familiar things . . . new in experience" also helps to illuminate features of the archaeological method under examination. Dewey suggests that every experience—whether with another human being or with an interesting cylindrical stone—is found in the interaction with a live creature and its particular environment. Such an experience, however, "only becomes conscious . . . when meanings enter it that are derived from prior experiences. Imagination is the only gateway through which these meanings can find their way into a present interaction." Since there is never a perfect match between current experience and past meanings, a twofold adjustment needs to be made by the imagination. For Dewey, the conscious adjustment of the new and the old is imagination. This process seems to aptly describe the work of any archaeologist whose goal is not simply to uncover material remains, but to explain their significance. The postprocessual approaches may be more liberal in their explanations, but they take their cues, if Dewey is right, from the imaginative dimension of cognition.

The leaps that the imagination makes between current experience and past meanings, like the theoretical leaps made by Hodder, Lewis-Williams, and Renfrew, are precarious and often open to ridicule. Dewey admits:

> There is always a gap between the here and now of direct interaction and the past interactions whose funded result constitutes the meanings with which we grasp and understand what is now occurring. Because of this gap, all conscious perception involves a risk; it is a venture into the unknown for as it assimilates the present to the past it also brings about some reconstruction of that past.[21]

Indeed, the so-called gaps—between the here-and-now and the past—that archaeologists face are formidable. In bridging these gaps and reconstructing the material, cultural, and intellectual relations of artifacts, cognitive archaeologists take risks in order to expose possible meanings that had hitherto remained hidden. Renfrew's development of *mappae* provides researchers a new way of interpreting old objects, as does Lewis-Williams' hypothesis of shamanism and its political motivations, and ways of suggesting analogues between the symbolic systems of antiquity and the systems of today. Hodder extends this appeal to novel interpretations, arguing metaphorically and analogically in his interpretation of grave sites. By comparing particular architectural aspects and drawing these aspects into a cohesive relation, he is able to imagine the symbolic significance of these relations. A new hypothesis strikes him, and he is able to imagine the thought that might have emerged in tandem with these particular material conditions.

It should come as no surprise that the archaeologist's interpretative approach violates Barnes' sensibilities, sensibilities that insist that objects-in-themselves reflect an obvious and fixed meaning. Dewey speaks to Barnes' discomfort by noting that "the theory of reality that defines the real in terms of fixed kinds is bound to regard all elements of novelty as accidental and esthetically irrelevant, even though they are practically unavoidable."[22] To say that postprocessualists' interpretations are imaginative is not to deride them as being imaginary, that is, accidental or irrelevant. The hypothesis that interpretative approaches propose—that material context may give rise to particular patterns of thought—can be tested and, in certain cases, falsified. Indeed, the analysis of Anaximander and the material conditions of thought should be regarded as such a testing.

In the next chapter, we will return to Dewey's understanding of philosophy and epistemology. His pragmatic conception of thought and inquiry gives credence to the belief that at least one central and previously unexplored aspect of Anaximander's thought can be best understood through an examination of the architectural and material context of his time; it can be understood by way of this approach, according to Dewey, because Anaximander, indeed all thinkers, *think through and by means of* their cultural and historical context. Before moving forward, however, I turn to examine the interpretation and translation of objects from the perspective of the late Anglo-American tradition in order to expose a type of crossover between the pragmatic and hermeneutic methods of Gadamer and Dewey and more recent developments of analytic thought, in particular, the thought of Quine, Davidson, and Putnam. These thinkers provide a way of framing the meaning of physical objects and a way of couching the work of both postprocessual and cognitive archaeology. Then, finally, I turn to Searle, whose recent arguments attempt to return to the classical path of metaphysical realism.

C

PHILOSOPHICAL STRATEGIES FOR
MAKING SENSE OF THE "REAL"

C.1

QUINE AND DAVIDSON:
THE INDETERMINACY OF TRANSLATION
AND RADICAL INTERPRETATION

In the "Two Dogmas of Empiricism" (1951), Quine presents an account of objects that he later develops in *Pursuit of Truth* and that seems to provide the epistemological underpinnings for archaeological interpretations of Hodder and other interpretative archaeologists. By 1953, Quine had revised the standard epistemic understanding of things that had gained traction in the early part of the 1930s and 1940s and proposes a position remarkably akin to those assumed in American pragmatism and continental hermeneutics. He writes:

> For my part I do, *qua* lay physicist, believe in physical objects and not Homer's gods; and I consider it a scientific error to believe otherwise. But in point of epistemological footing the physical objects and the gods enter our

conceptions only as cultural posits. The myth of physical objects is superior to most in that it has proved more efficacious than other myths as a device for working a manageable structure in the flux of experience.[23]

In saying that concepts of physical objects are only "cultural posits," Quine departs from the positivist stance that a physical thing can be reduced to an atomic, propositional, ahistorical meaning. The "myth of physical objects," the belief that there is a one-to-one correspondence between an object and its meaningful reference, may be "superior" to other myths, but it is a myth nonetheless.

Quine elaborates, suggesting that, to the extent that knowledge is mythical or man-made, there is always a choice involved in expressing particular truth statements. These truth statements make sense only through a consideration of their relations to the wider field of knowledge. Quine writes:

> The totality of our so-called knowledge and beliefs, from the most causal matters of geography and history to the profoundest laws of atomic physics or even pure mathematics and logic, is a man-made fabric which impinges on experience only at the edges. . . . But the total field is so underdetermined by its boundary conditions, experience, that there is much latitude of choice as to what statements to reevaluate in the light of a single experience. No particular experience is linked with any particular statement in the interior of the field, except indirectly through considerations of the equilibrium affecting the field as a whole.[24]

Here one must be careful not to overstate Quine's point. Quine does not say, like certain champions of indeterminacy, that the pursuit of meaning is bankrupt. He does not claim that it is impossible or impractical to state that this stone, this physical object, really is a column drum. He does, however, claim that such a statement can never be taken at "face value" as the sole meaning of the given object in the world. Indeed, the meaning of any utterance spoken in reference to an object of experience must always be understood in a wider system of meanings. This vision of knowledge as "holistic" resonated with pragmatists such as Dewey and sets the philosophical stage for scholars such as Hodder and Renfrew who aim to unearth the wider system of meanings by which to understand the reference to a particular object of history.

This epistemological holism, often referred to as the Duhem-Quine thesis, undermines the analytic conception of rationality in its implication that any translation made between an object of experience and an uttered reference must remain, in an important respect, indeterminate. Objects have discernible meaning only in a broader conceptual system, or "translation manual," that begins to explain the constitutive relations of an object. In the words of Quine, it is nonsense to ask *absolutely* whether the terms "rabbit" or "rabbit part" "really refer respectively to rabbits (or) rabbit parts. . . . We can only ask relative to some background language."[25] John Murphy notes that Quine believes that the meaning of objects is never exhibited as it might be in a museum—that is, divorced from the behavior and interaction of living individuals.[26] Meaning is never divorced from background. In this important and central sense, Quine echoes a type of pragmatic naturalism in his understanding of epistemology and ontology:

> When with Dewey we turn . . . toward a naturalistic view of language and a behavioral view of meaning, what we give up is not just the museum figure of speech. We give up assurance of determinacy. . . . When we recognize with Dewey that that 'meaning . . . is primarily a property of behavior,' we recognize that there are no meanings, nor likenesses, nor distinctions in meaning, save what are implicit in people's dispositions to overt behavior.[27]

At another important point in his lecture entitled "Ontological Relativity," Quine unpacks this claim. He suggests that the mind and knowledge "are part of the same world that they have to do with" and should be examined in the same empirical spirit.[28]

While Quine is often portrayed as a naturalist rather than a cultural relativist, Richard Rorty suggests that his work converges with the poststructural movement, the very same resource that inspired interpretive archaeologists and scholars from a variety of other disciplines. As Mary Hesse notes in her *Structure of Scientific Inference*, Quine either ignores or obscures the radical character of the indeterminacy of translation, often avoiding the way in which this understanding of translation necessitates a reevaluation of nature—but also history and culture—as important epistemological vectors.[29] The radicality of this move in the analytic tradition, however, is reflected in the work of Donald Davidson.

In 1967, Davidson published "Truth and Meaning." Heavily influenced by the work of Quine, he argues that any "theory for interpreting a speaker had to be tailored to a particular speaker and even relativized to a time."[30] Davidson echoes Quine in an important respect, stating that beliefs and meanings form an interconnected network and "are identified and described only within a dense pattern of other beliefs."[31] Davidson coins the term "radical interpretation" to refer to the process in which beliefs, meanings, and truths are made perspicuous. The obstacles faced in this type of interpretation, however, cannot be overcome through an appeal to a type of Quinean translation manual. In truth, according to Davidson, such manuals are often in scarce supply and are never exhaustive in bringing these relations to light. This fact is painfully apparent in both the fields of archaeology and ancient philosophy in which practitioners grapple with fragments of artifacts and texts without the help of a ready-made conceptual manual. Davidson wrestles with the difficulties of interpretation and arrives at a particular conclusion: Radical interpretation is always an open-ended process, indeterminate, anticipatory, and provisional.

Davidson's argument against the idea of conceptual schema is extended in detail in "On the Very Idea of a Conceptual Scheme." Here he revises Quine's conception of holism, suggesting that there exist no particular schemata that might be used to translate a complex world. He writes that "what we need is a theory of translation or interpretation that makes no assumptions about shared meanings, concepts and beliefs." This stance is too extreme for the kind of historical realism being defended here. This study is advocating the kind of realism where it is axiomatic that there are some shared meanings and constructs that both limit and enable our interpretations of artifacts as they show themselves to us. An archaeological approach to philosophy implies that there is a relation between a historical arti-*fact* and the meaningful and interpretative contexts in which these objects were used and then discovered. In the absence of common schema, Davidson opts for an approach to meaning and truth that resonates with an extreme form of hermeneutics. He writes: "We recognize that truth must somehow be related to the attitudes of rational creatures; this relation is now revealed as springing from the nature of interpersonal understanding . . . the conceptual underpinnings of interpretation is a theory of truth."[32] In order for us to interpret the meaning of a given proposition—such as "This is a column drum"—we must imagine ourselves placed in the particular community in which this utterance would make sense in an "interpersonal understanding." Such an imaginative move involves examining the necessary and sufficient conditions for understanding language and understanding meaningful objects in the world. These conditions could be formalized only to the extent that an interpreter was able to reconstruct the particular "state of affairs," the embodied and holistic context, of the utterance.

Davidson remains somewhat unclear on the particular method that brings about such a reconstruction. Perhaps this ambiguity is warranted, for he comments at points that such a reconstruction can take various forms. He does, however, suggest that any act of interpretation reflects what he calls the "principle of charity" or "rational accommodation." This principle cannot be described in any great detail prior to its use, but rather stands as a type of injunction to maximize agreement between an interpreter and the interpreted. Stated more simply, the principle of charity suggests that we regard interpreted systems of language and thought as being analogous—although not identical to—our own. Without the use of analogy and linking metaphor, the interpreter is barred access to the life and thought of seemingly foreign communities. In many respects, ancient Greece stands, in all of its cryptic pieces, as a foreign community to be interpreted. Cognitive and interpretative archaeologists seek to reconstruct the cultural and material framework in which the meaning of a particular object emerged. Davidson explains that the principle of charity is used to make these inferences and develop working hypotheses like the *mappae* hypothesis put forward by Renfrew. Unlike Renfrew, however, Davidson is primarily a philosopher of language, so the coherence theory of truth that he puts forward depends not on underlying relations that make our language meaningful in reference to the things of the world, but on the patterns and schemes implicit in the syntax and grammar of the language itself. The strength of this study, in contradistinction, depends upon the rejection of Davidson's extreme stance while still arguing for a meaningful sense of realism; of course, the account of Anaximander's thought and the origins of philosophy are open to debate and revision, but there are constraints to the narratives that archaeologists and scholars of ancient philosophy tell about the origins of philosophy. These constraints indicate the grounds for realism and point away from a Davidsonian extreme "radical interpretation." This study, then, advocates a coherence of meaning; it supports a type of internal realism that resonates more closely with the work of Hilary Putnam.

C.2

PUTNAM'S INTERNAL REALISM
OR HISTORICAL REALISM

We have addressed, on the one hand, the work of scholars of ancient philosophy, such as Barnes, and a range of archaeologists who believe that our understanding of artifacts reflects a one-to-one correspondence between the things themselves and our symbolic representations of these things. On the other hand, we have addressed thinkers who draw this stance into and open to question and finally, in the work of Davidson's radical interpretation, we have encountered those who reject this stance out of hand.

Hilary Putnam attempts to moderate Davidson's claim, advancing a view that he calls "internal realism."[33] Following in Kant's footsteps,[34] Putnam holds that the world we talk and know about is empirically real, but that it is dependent on the mind of the inquirer. Putnam writes that "'objects' do not exist independently of conceptual schemes. We cut up the world into objects when we introduce one or another scheme of description. Since the objects *and* the signs alike are *internal* to the scheme of description, it is possible to say what matches what." To say that objects are dependent on our conceptual schemes is neither to jeopardize an understanding of common sense nor to reject the idea that there are objects in the world, distinct from us, with which we maintain causal relations. Instead, to hold for internal realism is to maintain that the way in which the world is "cut up" is isomorophic with the way human conceptualization is "cut up." That is also to say that the world and the mental life of human beings are analytically but not ontologically separate. In Putnam's words, in the opening page of *The Many Faces of Realism*, "the mind and the world jointly make up the mind and the world."[35]

The question remains, however, as to how our conceptual schemes could access the world in any real way. Putnam states that "there are external facts, and we can say what they are. What we cannot say—because it makes no sense—is that the facts are independent of all conceptual choices."[36] The truth, therefore, is partially contingent upon the choices of an observer and partially dependent on the "facts" of the world. Thus, Putnam weds a particular type of conceptual relativism with a brand of realism. This approach will prove extremely important in the coming sections of this book, since it preserves interpretative pluralism, while not succumbing to the skepticism that attends radical constructivist models. Putnam, like Quine and Davidson, maintains that there is no true understanding of things-in-themselves, a tradition that rightly traces back to Kant. Unlike Quine and Davidson, however, Putnam maintains that we can still preserve a meaningful sense of "real truth"—to differing degrees—in each conceptual scheme that tries to approximate the reality of the world. This is made apparent in the second lecture of the *Many Faces of Realism* when he discusses the different paradigms of logic that might be held by different individuals.[37] Each of these logics is correct, and equally correct in his estimation, but remains open to testing in accord with the purposes of these individuals and novel environmental situations.

Having already addressed Dewey's understanding of the relationship between interpretation and truth, it should be obvious that Putnam's work is firmly situated in the pragmatic camp. Indeed, Putnam goes so far as to say that he should have called his internal realism "pragmatic realism" since the truth that informs a given conceptual scheme is partial, provisional, and pragmatic.[38] In sum, Putnam's point is this:[39] objects are individuated only relative to conceptual schemes, and conceptual schemes are selected by us relative to our purposes, interests, and values. Therefore, what "objects" there are, and what the "facts" are about any aspect of our experience, can only be defined relative to the values we have as inquirers and actors. There is no single way the world is in itself. Putnam simply rejects that side of Kant. But there is the world for us, and there are many possible "true" ways to carve that up, given our various purposes and values. There are some modest experiential constraints on that carving process, but none that selects out just one way to get the world right.

C.3

SEARLE'S EXTERNAL REALISM AND THE
SOCIAL CONSTRUCTION OF REALITY

John Searle, in *The Construction of Social Reality*, offers an argument for external realism, that is, an argument that claims to prove that there is a world that exists independent of any human agency. Moreover, the world that Searle argues for is not some Kantian noumenal world about which we can say nothing; there are "facts" about the world that he labels "brute facts," basic truths about the world that are describable in language. So, for Searle, unlike Putnam, it is possible to carve up the world one way to "get it right." The motivation to propose such an argument arises, as Searle explains it, given the philosophical scene in which it has become common to deny the existence of a reality independent of human representations and to deny that true statements correspond to facts.[40] To understand Searle's argument, we must see that he regards the two issues as inextricably connected. He argues for external realism and thinks that, having done so, "brute facts" follow analytically. But as we shall see, the core of the problem is not with his argument for external realism but rather with the brute facts that he regards as a necessary concomitant.

It is important to understand that what Searle means by external realism is not what Quine, Putnam, and others might mean by that term. Quine and Putnam, for example, would agree with Searle that there is a world not entirely of our making that stands outside of our conceptual schemes and languages. They would, however, not agree with Searle's commitment to the correspondence theory of truth that distinguishes their respective views. Searle holds that there is one correct articulation of how things are, ultimately resting on brute facts, and his critics doubt that we can make sense of any "facts" being "brute" because our descriptive languages emerge in cultural contexts, our theories always address questions produced in different historical circumstances, and because political and aesthetic considerations can never be separated fully from the practice of science. So, Quine and Putnam seem to fit the mold of Searle's antirealist—but that is because they reject brute facts, not external realism. As we continue, we should keep in mind that the question of brute facts must be answered as it is quintessential for our project. If Searle is right that there are brute facts, this would have profound implications for understanding the metaphysical presuppositions of the historical narrative I am advancing—indeed of any narrative or account.

Searle claims we can distinguish meaningfully between two kinds of facts: brute and institutional. Institutional facts are facts only through collective intentionality, and this means that they are socially constructed; these facts, however, are only possible in light of brute facts, that is, in light of some reality independent of human agency. Thus, the analysis of institutional reality—of the socially constructed world—reveals that there are things that cannot be so constructed—there are things that must exist independent of human intentionality. The structure of Searle's argument about brute facts is to begin by getting clear about institutional facts and to show that they bottom out, ultimately, on brute facts.[41] As such, Searle's argument is directed against two positions that he identifies as rejecting both external realism and brute facts: Searle directs his arguments against "phenomenal idealism" that holds that reality consists in conscious states, and "social constructionism" that holds that reality is socially constructed.[42] Searle argues that these anti-realist adherents must accept some version of "external realism," despite their protestations to the contrary. That is, these antirealists must accept some argument that there is over and above our systems of categorization some-thing that gets classified. And Searle regards such an admission as undermin-ing their antirealist positions.

Searle's argument is a transcendental one; he follows the kind of argu-ment that Kant proposes, namely, to assume that a certain condition holds and then to show the presuppositions of that condition.[43] The condition he assumes, following Wittgenstein, is that we do communicate successfully about our shared world, we make claims with truth values about things that are independent of us. According to Searle, in order for this communicative practice to be intelligible, a necessary condition for its possibility is that there is a way the world is in virtue of which we can say that something we describe is either true or false. Searle's argument can be summarized as follows:[44]

1. "The normal understanding of utterances in a public language requires that the utterances be understandable in the same way by any competent speaker or hearer of the language."
2. "A large class of utterances purport to make reference to phenomena that exist outside of, and independently of, the speaker, the hearer, and their representations, and indeed, in some cases, independently of all representations."
3. "Features 1 and 2 require that we understand the utterances of many of these sentences as having truth conditions that are independent of our representations. By purporting to make references to public phenomena, phenomena that are ontologically and not merely epistemically objective, we presuppose that the truth or falsity of the statements is fixed by how the world is, independently of how we represent it."
4. "But that presupposition amounts to the claim that there is a way things are that is independent of our representations, and that claim is just (one version of) external realism."

So, the transcendental argument is this: Only on the condition that there is a reality that exists independent of our representations could we communicate. And we do communicate, so there must be such an external reality. As Searle sums it up: "our ordinary linguistic practices presuppose external realism."[45]

This is all well and good for our purposes because in both the continen-tal and pragmatic schools that have been followed here, and among the mod-erate interpretative archaeologists, there is no doubt about the thesis of exter-nal realism. The crux of the problem for our project of understanding the metaphysical presuppositions of a new, different kind of historical narrative of (part of) Anaximander's philosophical mentality is to decide whether or not there are some basic facts that are describable in some primary, brute way. How can Searle accomplish this by a transcendental argument? In a transcendental argument, there is no epistemic gain; we can only clarify our conceptual situation by working out the conditions for the possibility of some x. In order for Searle's argument for external realism to do the work of illuminating brute facts, the argument must show the referential features of it. But, in all specific cases, external realism is nonreferential; we can take for granted that there is a world outside of and independent of us without accepting that any specific claim refers to any object in the world. When we accept Searle's notion of external realism, we accept the claim that our language refers to a world independent of our representations of it, but we are in no position by means of *this* argument to say, in fact, which and when. The trouble with Searle's argument is that it commits the fallacy of division; it mistakenly attributes the predicate "refers" to individual claims that it rightly attributes to the whole class: From his claim that all of our language is referential [part of external realism], he fallaciously infers that a particular claim actually refers to this or that thing. But this is what the arguments for brute facts must show, and this is what a transcendental argument can never provide because there is no epistemic gain, only conceptual clarification. The point is that something other than a transcendental argument is needed to make the case for brute facts. And Searle has not provided it.

Another way of getting to the same point is to see that we can accept external realism and deny brute facts without contradiction. Consider the problem this way. Since language use is a social process, successful communication requires that we must be able to refer to things. And our communication may very well rely on our supposition that the things we refer to are unchangeable. But the fact that we regard things as unchangeable does not make them so. Thus, our regarding some facts as "brute" does not make them so. The 'brute' is merely what we have no reason to challenge within a particular context of inquiry. However, it says nothing about the absolute character of some objective reality. It would be better to drop the term "brute" altogether, because it mistakenly ignores the crucial role of values, purposes, and contexts of inquiry. To put the matter differently, the claim that "language is referential" is not incompatible with the claim that "there are no brute facts." Let us pursue this issue further.

Searle's argument begins with institutional facts and reaches the conclusion that brute facts are presupposed by them. But there are two different formulations of "brute facts" operating in Searle's work.[46] On the one hand, there are "brute facts" where objects are language independent, such as "My dog has fleas," and "There is snow on top of Mt. Everest." We communicate successfully by means of such statements on the condition that there is a real world, ontologically objective; we are able to understand each other on the condition that our words refer to some publicly accessible reality that does not depend on your or my representations.[47] Searle's second formulation of "brute facts," that "x counts as y in context c" depends upon there being an x upon which agents can impose a status function. That is, institutional facts have the logical form of constitutive rules that rest ultimately on brute facts, a world of noninstitutional items that get understood in a particular manner. So Searle explains that "this piece of green paper (x) counts as money (y) when made in an appropriate way, by the appropriate agency," and so on; this rule might be reiterated many times, but it always bottoms out in brute facts.[48]

Let us distinguish, then, the two different formulations that seek to demonstrate brute facts by means of external realism:[49]

ER-1: The world consists of states of affairs, accurately describable in our current language, which do not depend on human agency in any way.

ER-2: In order to have a language, we must presuppose that there is a world, independent of human representation, that is capable of being conceptualized in myriad ways.

It seems that Searle begs the question of the antirealist by conflating two separate theses in his defense of external realism. This is because one can hold ER-2 and still subscribe to positions of the antirealists since it makes no substantive claims about the external world. ER-2 leaves us in much the same ontological position as Kant's defense of noumenalism; it is required as the external source of my representations, and yet about it we can say nothing. But throughout *Construction*, Searle repeatedly makes claims about the contents of the real world.

What then counts as a brute fact? Suppose I make the claim that archaic column drums exhibit *anathyrôsis* and *empolion* as part of the process of construction. This is certainly a claim about the empirical world; it asserts that a certain state of affairs is the case and, since it can be verified or falsified, has all the earmarks of an assertoric speech act, the claim can be understood. Searle insists that substantive claims about the real world require that this reality can be understood in more or less the same way by those making and hearing such claims. But rather than showing a robust reality, what it shows instead is that our assertoric claims about an external reality crucially depend on our antecedent understanding of this reality. "Before I can make any substantive claims about the nature of some 'x', it must be the case that we collectively understand 'x' in a particular manner—that it is recognizable by us as an object of predication."[50] Brute facts presuppose institutional facts, not the other way around. Or, better yet, all facts are institutional facts.

In the *Structure of Scientific Revolutions*,[51] Thomas Kuhn illuminated how thinking within a paradigm is made possible because certain predicates become deeply entrenched in institutional life, in what is at any time the prevailing worldview. The entrenchment of these predicates creates the sense that there are brute facts. But Kuhn helped to disabuse us of this false and misleading inference. The same kind of conclusion was reached by Nelson Goodman in his *Ways of Worldmaking*;[52] to assert that facts are theory-laden is not to deny external realism, to deny that there exists a world independent of our mental representations, but rather to affirm that there are versions of this world that correspond to different systems of descriptions. For Kuhn and Goodman, we start from one worldview and at times end in another; the appearance of rigidity and entrenched familiarity misleadingly creates support for the supposition that there are brute facts when it reflects only the entrenched institutional facts. Thus, Searle inverts the relation of dependence between brute and institutional facts: We can call some fact brute only because of entrenched institutional facts. The very existence of brute facts can itself be regarded as an institutional fact. This critique undermines Searle's position.

TEN

The Embodied Ground of Abstract and Speculative Thought

John Kaag, co-author

A

THE MATTER OF MIND: AN ARCHAEOLOGICAL APPROACH TO ANCIENT THOUGHT

HOW DID ANAXIMANDER think through his train of thoughts on cosmic architecture? He did so in the same way that we all come to understand so many things—by figuratively projecting one set of experiences that has become a stabile platform of understanding onto a set of experiences of a different kind that we are trying to understand. In his case, his embodied experience of watching architects work on the house of the cosmic power established a stabile platform of understanding from which he projected metaphorically onto the cosmos the stages and structures of its natural architecture.

To this point we have addressed the way in which archaeologists in general and cognitive archaeologists in particular infer particular systems of thought from objects of (pre)history. We have suggested that such inferences resonate with the concept of interpretation developed by the hermeneutic and pragmatic traditions and extended along particular analytic lines. These interpretations, however, cannot lead us back to the meaning of ancient philosophical thought unless it is shown that this thought was constructed by way of the material conditions that archaeologists now examine, albeit in somewhat dilapidated form. Stated more simply, it is necessary to investigate the extent to which a thinker such as Anaximander used his understanding of architecture and material objects in the building of his conceptual system. Only through such an examination does an archaeological approach to Anaximander seem warranted. This section will evaluate the assumption upon which the rest of the project relies: Anaximander's use of architectural and material terms should not be considered accidental or additive, but rather constitutive, in any exegesis of his philosophical thought.

To consider these terms as constitutive means to take seriously the way in which his thought, and thinking more generally, is embodied in a material context and is metaphorical in disposition. Can this case be made, and, if so, how? Recent research on the philosophy of mind, linguistics, and cognitive neuroscience has undertaken this project and shown that abstract concept formation relies heavily on the physical and material categories that emerge through embodied experience. In extending this line of thought, the aim is to show how this work, stemming from the sensibilities of twentieth-century pragmatism, might underscore the importance of metaphor in ancient philosophy. First, a brief recount will be provided of comments made by John Dewey and William James on the natural and material basis of mental life. Then our attention turns to more recent metaphor studies conducted by Mark Johnson and George Lakoff that attempt to describe the process of metaphor formation in greater detail.

B

JOHN DEWEY AND WILLIAM JAMES ON THE CONTEXT OF CONSCIOUSNESS

Dewey's work on interpretation helped situate the process of inference that interpretative and cognitive archaeologists have undertaken in recent years. His rendering of human consciousness and philosophic thought, however, can support the current project in another important respect. Dewey makes an argument for an assumption that underlies this project—the assumption that philosophy stems from the cultural particularities and embodied experiences of individuals situated in specific contexts. Dewey writes:

> Philosophy is seen to be a philosophy of history, not in the sense of explaining why history is or must be as it is, but in the sense that philosophies spring from and embody characteristic conditions and crises of human history. They are akin to religious, artistic or economic institutions and problems, to other forms of culture in short, and to a science as a historic mode of culture rather than to science in its abstract, non-historic sense.[2]

Philosophy is not apart from, but rather a part of, the cultural context of a society. In this sense, the truths that such a philosophy produces do not reside above us in some sort of eternal realm of ideas, but take place, very literally, on the ground of human experience. The questions that philosophers seek to answer are not developed ex nihilo, but rather framed in medias res, in the middle of the things of the world.

Dewey suggests that anthropology, in its concentration on the cultural and material realities of particular epochs, should be considered continuous with, although not identical to, philosophy. He expands on this point, writing: "Philosophy, even in its narrower and technical sense, gradually emerged from a background that no one can deny is the appropriate field of the anthropologist, while its whole course exhibits interaction with religious, scientific and political movements that fall within the view of the general historian."[3] Here it seems appropriate to extend this observation to the fields of architecture and archaeology. Archaeology frequently investigates the remains of an architectural context that provided the material relations that might have shaped the course of ancient thought. "The philosopher," Dewey states, "even in his cell or study, still derives his material and his issues from the currents of life about him."[4] These currents of life would include, among other events, major building projects in the area, developments in scientific and architectural methods, and novel cultural practices that would affect the construction of contemporaneous philosophic treatises.

This comment on the importance of the context of thought, expressed by Dewey in the 1930s, was anticipated by William James in his *Principles of Psychology* (1892). James addresses the way in which mental life is inextricably bound up with its material situation. James suggests that "mental facts cannot be properly studied apart from the physical environment of which they take cognizance." One of the greatest shortcomings of rational psychology—and here the rationalist epistemology suffers correspondingly—is its attempts to explain conscious activities "without reference to the peculiarities of the world with which these activities deal."[5] This understanding of mental states holds that ideas are permanent and can be grasped only through the use of rational and disembodied consciousness. James balks at this notion, writing that "a permanently existing idea of *Vorstellung* that makes its appearance before the footlights of consciousness as periodic intervals is as mythological an entity as the Jack of Spades."[6] If James is right, if humans do not think by way of ideal representations, the question of cognition still looms: How did Anaximander, or for that matter any ancient philosopher, think through a particular train of thought?

C

THINKING THROUGH METAPHOR
AND THE BODY OF KNOWLEDGE

The answer to this question about how Anaximander or anyone else thinks through a train of thought comes in fits and starts and remains less than fullfledged. It emerges from the work of American pragmatists of the early twentieth century, but also in the current work of philosophers and linguists who have spearheaded the discipline that has come to be known as metaphor studies. These individuals seek to expose the way in which cognition takes its

cues from the material relations of particular situations and uses these relations to intimate abstract content. In 1987, Mark Johnson, inspired in large
part by Immanuel Kant's notion of the schemata, proposed a metaphoric theory of cognition.[7] In the *Body in the Mind*, he suggests that the abstract meaning and highly complex rationality emerge from, and are limited by, the pat-
terned relations that organize the embodied lives of human beings.[8] This
emergence of meaning depends on the ability to think metaphorically, that
is, to figuratively project one set of experiences (in this case, the field of
embodied, physical forces) onto a set of experiences of a different kind. As
Johnson observes, the "metaphor consists in the projection of structure from
one domain onto another domain of a different kind."[9] This process of mapping is reminiscent of Peirce's comments on the character of abduction and
Kant's understanding of aesthetic judgment. Both depend on the process of
mapping disparate fields of sensory perception that harmonize—and harmonize even in the absence of pre-established concepts. Another way of saying
this is that the premises and domains of imaginative consciousness cannot be
pre-given or be exhaustively described. The mediating process that acts to
draw these seemingly separate maps into harmony has been described by
Johnson as being "imaginative" insofar as it is defined by a particular mediation, plasticity, variability, and spontaneity.

Johnson's interest in imaginative metaphor is not constrained to a particular study of aesthetics, but rather reflects a suspicion that aesthetics on
the whole may provide a valuable perspective in understanding the nature of
human cognition. In *Philosophy of the Flesh*, Johnson teams up with linguist
George Lakoff to elaborate on this suggestion. He writes: "Metaphor allows
conventional mental imagery from sensory domains to be used for domains of
subjective experience. For example, we may form an image of something
going by us or going over our heads (sensorimotor experience) when we fail
to understand something (subjective experience)."[10] In the construction of
abstract systems, individuals employ basic sensorimotor concepts that are literal as metaphors for nonliteral abstractions. Without these metaphors, such
abstractions are "relatively impoverished and have only a minimal skeletal
structure."[11] Anaximander's comment that the earth is like a "column drum,"
and, more literally, an organic "vertebra," begins to take on more significance
in the context of metaphor studies. This ancient metaphor might have been
a way to conceptualize an abstract moment of his philosophy. The metaphor
he uses does not reflect a particular propositional meaning but rather points
his listener back to the material relation in the world, and invites an audience to make a conceptual leap, a type of comparison, between the material
and the abstract realms.

This nonpropositional approach to language and cognition has been
advanced by Davidson, who suggests that metaphors, rather than meaning
something in particular, intimate something that encourages us to notice, see,

or re-cognize a relation in the world. He is very clear that "what we notice or see is not . . . propositional in character."[12] When the sheer ubiquity of metaphor is acknowledged in conjunction with Davidson's claims, it poses a formidable challenge to the standard propositional interpretation of cognition. Like any moment of radical interpretation, the metaphorical moment in which something is intimated cannot be reduced to a simple grammar that might correspond to simple truth values. Johnson and Lakoff agree, but insist that there must be an account of the way in which the metaphorical utterance used "is in any way connected up with what the hearer comes of notice" or re-cognizes.[13] The discipline of metaphor studies begins to develop such an account, explaining how metaphors are constructed and how they serve as heuristic devices used to sort out a world of ever-growing complexity. Archaeologists working on technological style, as discussed in Part I, chapter two, might find important insights in this literature.

Stuart Kauffman has an important contribution to make to this discussion although he is not a member of this metaphor studies camp. This scholar of theoretical biology and complexity theory, however, underlines the way that metaphor allows individuals to make sense of complex situations, noting that in the face of a biosphere of nearly infinite variation and complexity, embodied metaphor might be the only way for an organism to "get on with his business." He writes: "And metaphor? If we cannot deduce it all, if the biosphere's ramblings are richer than the algorithmic, then metaphor must be part of our cognitive capacity to guide action in the absence of deduction."[14] Indeed, more recent embodiment studies have indicated that metaphor is "a part of our cognitive capacity to guide action." In this regard, the work of Antonio Damasio has helped to show that a metaphoric imagination stems from basic modes of biological and social regulation. He writes:

> From an evolutionary perspective, the oldest decision-making device pertains to basic biological regulation; the next, to the personal and social realm; and the most recent, to a collection of abstract-symbolic operations under which we can find artistic and scientific reasoning . . . but although ages of evolution and dedicated neural systems may confer some independence to each of these reasoning/decision-making 'modules,' I suspect they are all interdependent. *When we witness signs of imagination and creativity in contemporary humans, we are probably witnessing the integrated operations of sundry combinations of these devices.*[15]

The sensibility that Damasio expresses—that complex processes of the human imagination have their basis in an interdependence and interaction of "lower" physical systems—has been expressed repeatedly by the scientific community for more than a century. Indeed, it seems fitting to recognize comments of American pragmatists—such as Peirce in his 1879 "Thinking as Cerebration"—as a preliminary step in this direction. Embodiment studies, however, only came into their own with the flourishing of phenomenology in Europe in the early twentieth century and only began to hit their stride in the late 1970s with the debates surrounding the field of cognitive linguistics. Even at this point, embodied and imaginative theories of language and cognition were considered the weak cousins of more analytically rigorous propositional accounts.[16] Only very recently have these theoretical underdogs been vindicated and recognized for their foresight in situating the embodied and metaphoric imagination at the core of human cognition.

The analysis of doxographical reports on Anaximander presented in this book rests on the suggestion made by Johnson and others that his thought arose in a particular embodied experience defined by the literal concepts of the architecture culture of the time. Only through an understanding of these literal concepts, explicated by archaeologists and historians, are we able to direct our attention back to the relations that Anaximander wished to intimate.

ELEVEN

Archaeology and Future Research in Ancient Philosophy

The Two Methods

IN PART I, I followed two "methods": a *Method of Discovery*, the process of discovery by means of which I came to formulate the new narrative about Anaximander, and subsequently a *Method of Exposition*, the argument form by means of which I attempted to justify the new narrative. Let us now turn to detail these methodologies.

A

THE METHOD OF DISCOVERY

The method of discovery describes the process by which we imagine the philosophical reflections of Anaximander or some other thinker witnessing the technologies that are alluded to in their preserved thoughts:

1. Examine the doxographical testimonies; see the hint(s) that suggests a connection to ancient technologies and techniques, and the temple building sites most especially.

   Or, more generally, examine the doxographical reports and identify references to ancient technologies, artifacts, or historical and cultural concerns.

2. Investigate what modern archaeologists and architects have informed us about how the architects worked, the details and problems of building, the surviving artifacts that constitute the evidence for their claims, and the

technological activities that were present at the building site. Provide whenever possible *illustrations* of those technologies.

Or, more generally, explore the evidence that archaeologists have provided for us about those technologies and techniques, the artifacts referred to, or historical and cultural context.

3. Connect the doxographical reports to these illustrated technologies by creating a new narrative account of Anaximander's fragments in this illuminated background of historical and cultural evidence. Whenever possible show how Anaximander's fragments can be illustrated and clarified, by appeal to the illustrated building technologies—the shape of the earth; the plan view of the distances to the stars, moon, and sun; the hollow-rimmed celestial wheels; the sun's fire that radiates and breathes as if from the nozzle of the bellows. The new narrative account invites us to imagine how, by means of building technologies, Anaximander came to see the cosmos in architectural terms.

Or, more generally, against this background of historical and cultural evidence, connect the doxographical reports to these illustrated technologies and techniques and/or the artifacts referred to in those reports.

B

THE METHOD OF EXPOSITION

The method of exposition follows from, and is a consequence of, the method of discovery. Once this process of discovery establishes abductively a series of working hypotheses, the narrative of these discovered hypotheses can be produced. Then, the next step is itself deductive to make the case. Accordingly the method of exposition has three parts:

(A) Set out the doxographical testimonies on particular themes.
(B) Set out the scholarly debates about the doxographical texts and the various interpretations emphasizing the contentious issues.
(C) Appeal to archaeological artifacts and reports to see if the long-standing debates can be resolved or clarified, and consequently help us deepen our understanding of Anaximander's belief system by illuminating the cultural context in which his thought developed and unfolded.

The method of exposition allows us to show that archaeology can be relevant to an understanding of the earliest philosophers, and thus the origins of philosophy traceable to the ancient Greeks.

The *method of exposition* needs no special defense. Its merits may be judged by the clarity with which the issues are presented and the compelling force of the archaeological evidence when applied to and connected with the doxographical reports. If that evidence is judged to illuminate the long-standing debates and, moreover, provide new insight to resolve or clarify them, then the method can be regarded as useful. The method of exposition, then, is an argument form adapted to reach from the archaeologist's resources to the philosopher's debates; it merely states the doxographical evidence, rehearses the scholarly debates, and then appeals to the archaeological resources to connect the historical and cultural context to the surviving fragments. The method of discovery is a different matter, and something more needs to be said here. The *method of discovery* is unfolded in the new historical narrative. The doxographical reports allow us to conclude that Anaximander identified the size and shape of the earth not merely with a flat cylinder but, more significantly, with a column drum. When we follow that reference to the building site, we come to discover that column diameter was the module for the architect's building of a cosmic house; his One over Many offered a modular technique by means of which the cosmic house was effectively built. And we know further from the art historical evidence that the column drum had symbolic meaning for the archaic Greeks contemporaneous with its introduction into monumental building. When we delineate a variety of techniques and technologies, observable at the building site, according to archaeologists, and then apply them to a variety of debates about Anaximander's surviving fragments, a new and deeper picture develops about what he thought and how he thought about it, that is, within his historical and cultural context. The merits of the new narrative, of course, may be judged by the series of studies that resulted from this method.

If the arguments enumerated in Part I, following the method of exposition, are plausible, and taken together compelling, then *why* is this so? The series of studies in Part I followed from the method of discovery. That method placed Anaximander at the building site to watch the architects at work. But how shall we justify, even if our supposition is correct, our thinking that "we understand" what Anaximander "would have seen"? The method of discovery requires that while we imagine what Anaximander was witnessing, we proceed by supposing that *there is some meaningful sense* in which the archaeologist, the "historian of philosophy," and Anaximander himself, are all "witnessing" the *same* projects at the building site.

We will never bring back Anaximander to reply to our suppositions about him, and there are good reasons to be doubtful about our abilities to recapture precisely what meaning monumental temple building—or for that matter anything at all—meant to these ancient peoples. This, I believe, is the best we can hope for: to produce cogent narratives that offer a coherent views that fits together the many puzzle pieces we have about the early philosophers in their historical and cultural context, subject to further review and revision.

TWELVE

The Application of Archaeology to Ancient Philosophy

Metaphysical Foundations and Historical Narratives

———————

IN THIS BOOK, I am providing an historical narrative about Anaximander of Miletos and consequently part of the story of the origins of philosophy connected with him. A central point is that the *historical* and *cultural* dimensions of this narrative prove to be philosophically illuminating. By this I mean that we gain insight into Anaximander's abstract and speculative thought process by tracing out his lived experience, re-created by aid of archaeological resources. Situating Anaximander's philosophizing within the narrative of temple building provides for a different kind of realistic narrative. Accordingly, I turn first to reflect on the metaphysical issues surrounding the kind of realism that I am advocating in this narrative, and subsequently reflect on how historical considerations fit within those metaphysical foundations.

A

THE *REALISM* IN NARRATIVE ACCOUNTS

A narrative account refers either to a "real" world that exists external to my conscious states, or it does not. Is there anyone who does not think that there is a world external to us? Well, the solipsist holds such an extreme position, claiming that all we experience is our conscious states, and insisting further that there is no ground for inferring that the contents of my consciousness represent anything else outside of me. The solipsist contends that he or she is the whole of reality and that the external world and other persons are representations of that self and have no independent existence. Kant's argument against such a position, pivotal in the history of philosophy, was that a careful inspection and inventory of the contents of consciousness reveal an ingredient that cannot be reduced to other conscious states. He called it "the given." In his analysis of cognition, he reached the conclusion that we would never have self-consciousness, indeed consciousness at all, if we did not receive sensations from "the given," outside of and external to ourselves. The solipsist is wrong, in Kant's assessment, because were it not for the existence of objects external to us, we would never become self-conscious. But we are conscious of our self. So the solipsist has it wrong. The solipsist's position is self-stultifying. For Searle, the solipsist is also wrong, but for a very different reason. For the solipsist to articulate his position, he must admit that the meanings of particular terms in a language are public, and the admission of a *public* directly undermines the solipsist's position. For if the meanings were not public, they would have to be private, to him alone, but for the solipsist to express his position he must presume public meanings; the admission of a public, outside of himself, undermines the claim that there is nothing other than myself. Thus, solipsism, again, turns out to be self-stultifying: one defeats one's own thesis in the process of declaring it. For Searle, solipsism turns out to be a performative contradiction; the very stating of the position is undermined by the condition for the possibility of meaningful assertions. So, who does not believe there is a world external to and independent of us? No one. In trying to get clear about the metaphysical presuppositions of this historical narrative, my account is clearly *realistic*. The matter is to explain the *kind* of *realism* it is. In what follows, I propose to sketch an account of what I am calling "experiential realism," but, to get there, I first turn to explore further the historical turning point for philosophers in rejecting metaphysical realism.

B

THE HOPELESSNESS OF METAPHYSICAL REALISM

Kant's *Critique of Pure Reason* sounded the death knell of the ambitions of metaphysical realism. In the conventional recitation of the metaphysical programs of Plato and Aristotle, it is supposed that there is a reality, singular in nature, and our epistemological task is to achieve clarity in our understanding of and our narratives about the world to adequately and accurately express that metaphysical reality. To adapt the language of a children's game played for generations, the intersection of our epistemological and metaphysical tasks is to "pin the tail on the real." There is a reality out there, independent of my inner states, independent of my consciousness, and the philosophical task of that program is to move beyond the shortcomings that are emblematic of them in order to grasp what *is*, what genuinely corresponds to them and is the source of these mental states. After all, why would I have the *spe-*

cific mental contents that I do—the awareness of walnut trees on my lawn, for example—if they were not caused by something outside of me, *similar* to them? On this approach, knowledge consists in my mental contents *corresponding* to them. Here we have an overview of the traditional program of metaphysical realism that brings together the "similitude theory of reference" and the "correspondence theory of truth."[2] There is a reality external to me, and, with philosophical instruction, I can come to know it; it is in this sense that epistemology dovetails with metaphysics. Kant showed in the *Critique* that this project of metaphysical realism was hopelessly flawed.

Kant's argument was that the structures of our consciousness brought conditions to our knowing that would forever prevent us from coming into contact with "the real" that nevertheless was the source of those mental representations. Kant's meaningful legacy in this domain consisted of his reminding us that, in the process of cognition, we impose epistemic conditions on the experience and so can never grasp the real independent of these conditions. Kant regarded these conditions to include space and time, and a twelvefold categorial structure of cognition. Thus, Kant held that the structures of our cognitive apparatus condition the thing that we claim to "know" and thus limit our possible knowledge of those things themselves; these structures that make knowing possible at once prohibit us from gaining knowledge of any thing independent of our conditioning apparatus. Let me try to elucidate this point by analogy. Let's imagine that we are wearing green sunglasses, and let us suppose further that we cannot take off the glasses and, moreover, that they cover our eyes so completely that we cannot see above, below, or around the green glasses. Now, let us suppose that Kant asks us: What is the color of the objects we see? The answer must be "green." Insofar as an object can be an object of experience for me, it must be (i.e., appear) green. But since we know that the green glasses condition my experience, and that this is how objects must appear to be an object of *my* awareness, we now ask the corollary question: What is the color of the object *independent of* my green glasses? Kant's *Critique* showed that the question was forever unanswerable, and yet, from the standpoint of our human reasoning nature, would never go away. We cannot stop asking about the nature of things as they are in-themselves, the source of all my mental representations, because of the very nature of reason itself that seeks to know the unconditioned; yet since such knowledge would lie beyond my capacities, properly understood, such questions would always remain unanswerable. Because we can never get outside of our cognitive structures, while we can grasp that things-in-themselves must be different from the way they "appear" to us, we can never give content to the details of those differences. Thus, Kant showed that the traditional project of metaphysical realism that sought to match epistemic clarity with the things-in-themselves was hopelessly flawed. We could know only the limits of our cognition—that is, how an object must be in order to be an object of experience for me.

What Kant achieved in his first *Critique*, then, was a rejection of metaphysical realism at a price. The old program, as he saw it, rested on the similitude theory of reference linked inextricably to the correspondence theory of truth. Once he showed that the similitude theory of reference was indefensible, the correspondence theory of truth fell along with it. The way the world was in-itself was inaccessible to us and unknowable by us, and thus there could be no objective knowledge of the way things are unconditionally, independent of our cognitive structures. Thus, Kant pointed us away from the project of metaphysical realism. Kant believed that we could still have "objective," indeed absolute, knowledge, but now confined to the structures of subjectivity—subjective conditions we imposed on experience to make it knowable—and this approach allowed him to focus on the conditions under which all sensible intuition was received and the categorical structures of our understanding. The result was that Kant urged us to reject the possibility of metaphysical realism; instead he advocated for "empirical realism" *and* "transcendental idealism." What this means, in his formulation, is that space and time were conditions that our sensible intuition imposed on experience so that insofar as we experienced anything as "real"—since all knowledge for Kant begins with "experience" (i.e., sensible intuition)—things appear "side-by-side" (space) and "in sequence" (time). But these are conditions of all of our knowledge, this is how we experience, and it is in this sense that we have knowledge of the "real"—empirically real. All of our experience and indeed theoretical knowledge is grounded in spatiotemporal objects. But since space and time (and the categorial structure of our cognition) are what we bring to and impose on the experience, things so far as they "appear" to us are not as they are "in-themselves." That is, things-themselves are *not* in space and time, but since this is the only way we can have experience of them, we can never know them unconditionally—transcendentally ideal.

With Kant's writing of the *Critique*, the age-old problem of metaphysical realism was undermined, whatever we come to think of the viability of empirical realism and transcendental idealism. And I am not about to defend Kant's project, but rather call on him as a decisive transition point in the history of philosophy. What Kant got us to see was that in the process of knowing, and the conditioning of the experience, we made it impossible to ever get at the thing-unconditioned, the thing-in-itself, the solution to the metaphysical game "pin the tail on the real." What it left us with, according to me but not to Kant, was to search for a meaningful sense of knowledge against the varying horizons of the purposes and values that direct our inquiries. As I have written elsewhere, Kant was right to reject metaphysical realism, that the locus of the real could be found in the thing itself, outside of and independent of me, but he was mistaken to embrace the project that absolute certainty was still to be found, this time in the structure of subjectivity.[3] Because Kant rejected the project of seeking objectivity in the object-itself, he reinterpreted the project of seeking objectivity in knowledge along the lines of his empirical realism but transcendental idealism. "Objectivity" turned out to be "rule-directed synthesis," the employment of the faculty of understanding by means of which something could become an "object" of experience for me. So, without embracing his position but following his lead in rejecting metaphysical realism, let us turn to consider the metaphysical presuppositions of the narrative I am proposing. Grounding the new narrative requires a clarification of the meaning of "objectivity" when absolute certainty is rejected *both* in the object and for the subject. Accordingly, I turn to summarize the argument of Part II and show further how the appropriate meanings of "objectivity," "fact," and "interpretation" reveal and ground the metaphysical foundations of our historical narrative.

C

CRAFTING A CASE FOR "EXPERIENTIAL REALISM": THE ARGUMENT OF PART II

The argument of Part II began with the claim that archaeology has been systematically and routinely neglected as a resource for understanding ancient philosophy because of a dominant, analytic approach to the business of philosophy that has regarded historical, cultural, and material matters to be nonstarters—as Barnes put it "philosophy lives a supracelestial life beyond the confines of time and space." In that supracelestial vision, reality has a singular nature, and its challenge is articulating metaphysical realism. Interpretative archaeology, along with hermeneutic and pragmatic interpretative approaches in philosophy, have shown us how we can ground historical narratives while rejecting the project of metaphysical realism, that is, without accepting that reality is singular and knowledge consists in our narratives that *correspond* in some one-to-one manner to that reality. When we do appeal to archaeological resources—architectural and material issues—we must see that these historical and cultural reflections are neither accidental nor supplementary to Anaximander's philosophical thought but rather constitutive of it. Philosophy emerges from our material embeddedness, can never be fully extricated from the historical and cultural moorings in which it is situated, and is metaphorical in disposition: Our philosophical understanding is an imaginative projection that speculates and abstracts from our embodied experience. The metaphysical project deserves to be called, for want of a better expression, *experiential realism*: there is an external world independent of our conscious states, but depending on our *lived experience* and *grounded experience* and the inquiries, values, and purposes that direct them, the world resists a single description. We can speak meaningfully about "objectivity" so long as we are clear that objectivity is about standards of epistemic assessment, not about the nature of things, as the failed program of metaphysical realism pursued, and it is and must be relative to values and purposes of inquiry. Hence, "objective knowledge" must always be regarded as an hypothesis tested through ongoing inquiry, and not simply a completed result of it. Our hypotheses about Anaximander or any thinker can, of course, be shattered by experience because the meaning of any x (experience, event, symbol, word) is what it points to by way of experience—past, present, or anticipatory. Meaning is never fixed for all time, but rather shared within a culture, that is, within historical and cultural contexts. Thus, we should avoid advancing claims as if they were in principal irrevisable, transhistorical, transtemporal; there are no such things as "facts" in this sense. Rather, a "fact" is a kind of stabile posit relative to a set of descriptions; it has a place within a web of concerns, within a context, within a mode of inquiry. The so-called fact becomes a posit around which things hinge, cluster, or hang together. Thus, my case is that the dichotomy between "fact" versus "interpretation" is a false and misleading one. It is not as if there are certain facts of the matter and interpretations are spun from them but rather that "facts" are starting points relative to values and purposes of inquiry; in this new narrative about Anaximander, "facts" are already inextricably interwoven with/in interpretations and they stand to rise or fall when subjected to challenge, testing, and debate. Let me attempt to clarify one more time the nature of the historical narrative by analogy to an imaginative jigsaw puzzle.

Let us imagine ourselves confronted by a myriad of puzzle pieces and no picture of the "finished puzzle" is furnished.[4] Moreover, let's imagine a puzzle whose pieces can be fitted together consistently in more than one way, creating a different "big picture" in the process. And let us imagine further that while the pieces can be placed together in more than one way, they most certainly cannot be fitted together, one and all of the pieces, any which way. This imagined puzzle bears analogy to our study of Anaximander and the new narrative I am presenting over and against older accounts. For in the collection of puzzle pieces before us, the parts do not determine the pattern, but rather the pattern determines the parts. In isolation, a piece of the puzzle means nothing, but, when assembled, the pieces come to acquire a meaning. Each piece counts as a "fact," but only with regard to one of the "big pictures" that can be produced by means of it. As the process of connecting pieces continues, and a pattern in the picture emerges as a whole, the meaning of each piece becomes intelligible. The assembly of the puzzle pieces that make a coherent picture is the assembly of warranted beliefs. If all the pieces can be placed together without internal contradiction, then those pieces overcome the recalcitrance of a series of beliefs. Of course, it does not follow that the assembly of puzzle pieces is the only way they can be connected and still make sense, but such is the defense of any coherent picture we can produce. Let us now turn to re-collect the puzzle pieces of Part II of this book.

In chapter seven, I drew attention to the point that archaeology has rarely been appealed to, if ever, to explain ancient Greek philosophy, and apparently not in any full-length study other than my own. I noted collaterally that ancient philosophy is one of the few disciplines that explore the ancient Greek world without regarding it of disciplinary significance to actually go to Greece. And I took these admissions to point to a certain conception of the business of philosophy that, despite all the important work that has stemmed from the conventional approaches, is too narrow to adequately capture a relevant and important part of it, and certainly its *origins* before the role of "philosopher" was named nonetheless secured. Unless scholars are prepared to return to the hypothesis that Greek philosophy began ex nihilo in ancient Greece by way of the "Greek miracle" suggested by learned scholars such as Burnet, Heath, and others at the beginning of the twentieth century,[5] then an historical narrative of the originating stages of philosophy must attend to transitions within the cultural context of archaic Greece. Archaeology can help furnish this narrative, but not merely because it incidentally supplies evidence for the cultural situatedness of the earliest philosophers but also because at every level of philosophical development, abstract and speculative thinking is fueled by patterns of life emerging in historical contexts. Lived experience is the ground for philosophical thinking, and archaeology can provide a window into its origins.

The next step in the argument, chapter eight, was to explore theoretical underpinnings when archaeologists construct their own narratives with specific focus on how abstract ideas are inferred from artifacts. How is it that a column drum is constructed, and, moreover, could suggest by some analogy the shape of the earth? What can we infer about a wheeled vehicle based on a surviving axle, or of a toy model and, moreover, how can the form of the sun bear a relation to a wheel? What does a vase painting suggest in its depiction of a bellows, and how can we infer that the cosmos is alive by breathing by means of its mechanism? This exegesis offered an opportunity to review changing approaches in archaeological theory—old, new/processual, postprocessual, and cognitive—and those changes also reflected epistemological issues about the *reality* of the past. One of the constructive results of that chapter was to give voice to postprocessual/interpretative approaches. They hold that reality is plural, and artifacts are multiplicities; the meaning of "real" will depend on the context in which it is used, with regard to the work done on and with the artifacts. The theoretical mistake of the positivistic processualists is to suppose that there can be a "real" past in the *singular* sense, some ultimate frame of things onto which our narratives can be mapped definitively and to which they correspond. The reality for which the interpretative archaeologists argue, then, is not an absolute any more than the objectivity that they seek. "Reality" and "objectivity" are not absolute or abstract qualities. The account that best resists the challenges of other competing theories, and of evidence, is the one that enjoys the designation of "objectively better." But when and if new evidence and new challenges arise, the archaeologist's narrative survives or falls, and with it the series of interconnected claims. When a narrative is challenged, interpretative archaeologists recommend the excavation of another site, the mobilization of another army of people, heterogeneous mixtures of peoples and things, to see if the new results challenge or lend support to earlier claims, to see if it stands up to a diverse challenge from a range of multiple perspectives. This is the best we can hope for when we produce our narratives by appeal to archaeological resources.

The next step in the argument, chapter nine, was to review mainstream philosophical approaches that would reveal epistemological and metaphysical foundations that were sufficiently broad to incorporate and ground the archaeological approach that guided the new "Method of Discovery" proposed here. By means of archaeological resources, we re-created in Part I of this book the preparation and installation of column drums, the techniques of plan-view imagination, the hollow-rimmed wheels on vehicles of transport, the bellows that mimicked cosmic breathing, and a reconstruction of a seasonal sundial. Our method of discovery consisted in *imagining* it through, again by means of the archaeologist's narrative; we *discovered* central features in the embodied experience of watching temple building, and we "shared the experience" with Anaximander—the Anaximander whose belief system we know from generations of scholarship. The abstract and speculative thought under investigation, then, is the imaginative projection of Anaximander's reflections on techniques at the temple building sites, literally *housing* the abstract cosmic powers, to illuminate the cosmic structure and mechanism of our house that *is* the cosmos. The architect's task was to house the idol of the god and to do so in a manner symbolically fitting of the cosmic power. The temple was a cosmic embodiment, a metaphorical transformation and reification of the abstract cosmic power of the god into a material, concrete representation, appropriate and befitting; Anaximander looked to the cosmos as itself an embodiment, grasped in a quasibiological way that, as a living creature, it had not started out the way it now appeared, and wrote the first philosophical book in prose that began plausibly by explaining how it came to be what it now was, a definite and determinate thing, out of the *apeiron*—an indefinite and indeterminate stuff. Anaximander's book articulated the stages by which the structureless became structured, and how that structure was built and by means of what mechanisms, appealing to building techniques he learned from the architects and which his audience of Ionians would have witnessed firsthand. With this narrative in mind, we explored two interpretative approaches, one hermeneutic and the other pragmatic. The hermeneutic approach opened our way to the imaginative meanings of objects and the pragmatic approach drew our attention from material context to imagined thought. What these approaches shared was a meaning of objectivity that resonated with interpretive archaeology; reality is plural, not singular, and objectivity is context dependent. Our understanding begins with bodily experience, our encounters with objects, and, depending on the nature and direction of our inquiries and purposes, abstract thought appears as a metaphorical projection from our material context and engagement with objects. To flesh out the epistemological and metaphysical grounding for this project, I turned to consider some debates about realism in the work of Quine, Davidson, and Putnam, and contrasted these sharply with Searle. Searle is the champion of external realism, his version of metaphysical realism. In his version of this project he seeks to identify (a) an external world independent of our conscious experience that is (b) describable ultimately by a series of *brute facts* that correspond to that world, and distinguishable from institutional facts that are social constructions. Searle's two-pronged argument relies on our use of language, that our words have public meanings, to make the case for the existence of an external world. And that part of the argument is fine. But he is unable to argue convincingly for the corollary claim of "brute" facts that are quintessential to his case for external or metaphysical realism. As Quine and Putnam, Kuhn and Goodman have argued in their own ways, there will be no way to succeed at this project. There are versions of this world that correspond to different systems of descriptions; within each system of descriptions, some "facts" appear "brute," but, from this appearance of entrenchment and solidity, it does not follow that they are brute in the required, primary, nonsocially constructed sense. There can be no single way to express a fact as brute—across different systems of descriptions, in different historical times, within and across different cultures; with no single way of describing them, Searle's argument for metaphysical realism falls. For Kuhn and Goodman, on the contrary, we start from one worldview or paradigm and at times end in another; the appearance of rigidity and entrenched familiarity are consequences of the prevailing paradigm or worldview, and they misleadingly create support for the supposition that there are brute facts when they reflect only entrenched institutional facts. Searle's brute facts turn out not to be what bottoms out after exploring socially constructed institutional facts, as his project requires; instead, what he calls brute facts prove to bottom out as other institutional facts. Thus, the problem with metaphysical realism is not its presupposition that there is an external world independent of us; everyone (almost) agrees. The problem is trying to show that there are basic truths, describable in language, transcending historical, cultural, and material contexts corresponding to that external world. Properly understood, this project of metaphysical realism is as wrongheaded as it is hopeless.

Next, we turned in chapter ten to get clear about how our bodily experience grounds knowledge. How did Anaximander think through his train of thoughts on cosmic architecture? He did so in the same way that we all do—by figuratively projecting one set of experiences that has become a stabile platform of understanding onto a set of experiences of a different kind that we are trying to understand. By Johnson's assessment, cognitive understanding can be explained as fundamentally metaphorical; this is a way of expressing how conventional mental imagery from sensory domains is projected imaginatively onto other domains. This process is itself abstraction. Our abstract and speculative thought, a central part of the business of philosophy, is the imaginative projection of our sensorimotor experience; the lived experience of watching the planning of temple architecture, the quarrying, transporting, finishing, and installing of megalithic masonry was projected onto the cosmic architecture of stars, moon, and sun. This new narrative about Anaximander, then, offers to address a systematic myopia about investigating philosophical thought in its historical and cultural context.

Finally, in Chapter eleven, we set out the two methods that were employed in this study. The first part was the method of discovery and this approach has no reception history[6] in secondary scholarship on the pre-Socratics. The originality and persuasiveness of this study rest largely on the development and employment of just this method of discovery. Noting the technical terms and analogies in the doxographical reports, we followed them where they pointed—to ancient technologies and techniques themselves; we found them at the ancient temple building sites where, thanks to archaeological resources, we were able to imagine some of the many activities taking place there. Then, we connected those archaeological narratives depicting lived experiences to the surviving doxographical reports to clarify and illuminate them. The second part was the method of exposition, and this was far more familiar and straightforward; it supplies an argument form to make the case for the usefulness and appropriateness of the method of discovery. The method of exposition consists of three parts: the presentation of doxographical reports, the debates about what the reports mean, and finally the appeal to archaeological resources to clarify and illuminate those long-running scholarly debates about ancient philosophical thought.

D

THE PRESENCE OF THE PAST AND THE
PROBLEM OF THE SUPRACELESTIAL THESIS

We began Part II of this book by critically focusing on Jonathan Barnes' supracelestial thesis that "philosophy lives a supracelestial life beyond the confines of time and space." How curious it must seem to a reader not specializing in this field that a book about ancient philosophy, about philosophy in an ancient period of history, should pronounce that historical considerations are irrelevant to such a study. Yet this antihistorical pronouncement represents an entrenched, even dominant, analytic approach to the study of ancient philosophy. Our study, in contradistinction, has tried to show that historical and cultural context can be very much relevant to our understanding of Anaximander's thoughts and part of the story of the origins of philosophy.

We have taken the position that historical knowledge is no different from any other kind of knowledge. When we claim that something is an historical fact we mean to say that—like any other fact—certain claims should follow from it, or be connected with it. But, unlike Barnes, when we ask what *kinds* of claims should follow from or be connected with any purported fact, we cannot resolve the question by appeal to supracelestiality. Historical inquiry is required to establish more precisely just what a particular point of view is; only when we explore a claim in its historical encounter can we become clearer about how a given point of view establishes or fails to establish its rational superiority relative to its particular rivals in some specific context. Thus, our claim that it is a fact that Anaximander was influenced by architectural practices of his day is one from which certain claims should follow. This means, for example, that once we understand how the architects imagined in both plan and elevation views, and that Anaximander also imagined the cosmic numbers using the "plan" technique, the long-running scholarly debates about what exactly are the cosmic numbers, and how a non-spherical, cylindrical earth can be in equilibrium from cosmic extremes, gain new intelligibility. All we ever have to make our case is the related evidence we can bring to bear on it; that evidence can be historically, culturally, and materially embedded.

An historical narrative is an assembly of "facts" and, as we have considered, those facts are posits, beginning points, selected for a narrative that reflects specific purposes and values of an inquiry. Those "facts" are connected by means of a causal narrative where the "facts" selected and the "causes" developed are bound to the historian's cultural context and the questions he or she takes as guidelines for investigations.[7] Dewey made just this point that all historical narratives are constrained by contextual demands and, moreover, that the contextual demands are set by the historical period in which they are written.

> The slightest reflection shows that the conceptual material employed in historical writing is that of the period in which it is written. There is no material available for leading principles and hypotheses save that of the historic present. As culture changes, the conceptions that are dominant in that culture change. Of necessity, new points for viewing, appraising and ordering data arise. History is then rewritten. Material that had formerly been passed by, offers itself as data because the new conceptions propose new problems for solution, requiring new factual material for statement and test.[8]

Our commitment to this approach consequently dismisses the idea of certainty as a transcendent standard for judging historical narratives.[9] To adopt the position that the historian selects a particular point of view, for any number of reasons, is to reject the correspondence theory of truth as a non-starter. The success of an historical narrative is not to "pin the tail on the real"; this is the wrong game. There is no God's eye point of view from which the story of Anaximander's thoughts or, more broadly, the origins of philosophy can be told.[10]

In part of our narrative in Part I, Anaximander's cosmic numbers are debated and reconsidered. The cosmic numbers are associated with the fiery wheels of sun, moon, and stars and are woven into a narrative. We have doxographical reports that identify the sun wheel with the numbers 27 and 28, and the only number associated with the moon wheel is 19. We have no testimony for the number(s) of the star wheel, though we have the report that the stars are closer to us than the moon, while the sun is most distant. And we have a report that the sun is the size of the earth. These reports may all be considered as "facts"—starting points—to be investigated, tested, rejected when necessary on evidence of inconsistency, and connected in a variety of ways. Now, once we accept that a singular causal statement—here, that architectural technologies connected with monumental temple building informed the way in which Anaximander imagined the cosmos because he envisioned it as cosmic architecture—depends on the viewpoint from which any historian selects one contributory cause rather than another, and once we accept that the historian's interest may determine both the point from which the narrative starts and the point at which the narrative ends, along with other entrenched milestones that are selected in chronicling the subject, we see that neither history nor that chronicle may be called a map, a copy, a picture, or a mirror of some antecedent reality.[11] To adopt this approach is to undermine the traditional project of the correspondence theory of truth. Historical narratives, such as this one about the origins of philosophy, then, are cultural products that respond to and echo present problems. The claims in an historical narrative can only be meaningful in a wider cultural context; no claims can have meaning in isolation. Lacking some external determining structure, the components of history—the "facts" included and the narrative that links them together causally—are a product of the historian's interest in a particular cultural context.

Of course, the writer of an historical narrative wants to produce "accurate" statements about the past. But the "past" is not a thing to be "reified" and our "accuracy" cannot mean "pin the tail on the *real* past," as if it existed somewhere apart from the present. The past, like the anticipatory future, exists meaningfully only in the present. So, George Herbert Mead had it just right when he writes:

> we can say that the only pasts and futures of which we are cognizant arise in human experience. They have the extreme variability which attaches to human undertakings. Every generation rewrites its history—and its history

is the only history it has of the world. While scientific data maintain a certain uniformity within these histories, so that we can identify them as data, their meaning is dependent upon the structure of the history as each generation writes it.[12]

What this approach suggests, and which our approach shares, is that the past and present are connected and united, not divided.[13] The past can only be a part of the present; it exists nowhere but in the reception history and the cultural memory of a society—in the history culture. Reception history traces the meanings that have been imputed to archaeological monuments and artifacts, and historical events; appeals to cultural memory invoke ways to consider the underlying agendas, processes, and social predispositions by which historical narratives are selected, "approved," and disseminated. For our study of Anaximander, reception history and cultural memory apply in a concerted manner; a reflection on both shows us that the narrative offered in this book about Anaximander and the origins of philosophy has not been considered heretofore in part because the ideals governing current scholarship in ancient philosophy have not regarded matters of historical and cultural context as a suitable source for unfolding the abstract and speculative ideas that are central to it. The reception history consists in scholarly literature and commentaries produced over the centuries, and more recently in course texts by means of which the ideas and place of Anaximander in the history of philosophy have been disseminated. There is no evidence of our narrative in that reception history. A deep, underlying question that this study has addressed is why archaeology is routinely and systematically passed over in mainstream studies on ancient philosophy. It is the centrality of the "supracelestial thesis" that lies at the crux of the problem.

Thus, it seems fitting in closing this study to return critically to Barnes' guiding principle that "philosophy lives a supracelestial life beyond the confines of time and space." Our study has not intended to undermine generations of scholarship that have proceeded with Barnes' supracelestial thesis in mind, whether or not the writers actually made such a declaration. But alongside these familiar approaches, studies that take seriously historical and cultural context deserve to have a place. Who would have thought that from an examination of ancient wheeled vehicles, the blacksmith operating the bellows at the forge, a review of architectural planning and building techniques—thanks to archaeology—we would gain new insights into Anaximander's ideas about the cosmos and the processes by which he came to those ideas? The usefulness and relevance of archaeological evidence are testimony to the usefulness and relevance of historical and cultural context for understanding the early stages of philosophical speculation. Moreover, it is time to acknowledge that the supracelestial thesis is also an agenda-driven part of cultural memory, with a long reception history. In the end, we must come to see the supracelestial thesis for what it is, despite protestations from its advocates, for there is no God's eye point of view available to us mortals; it is only another kind of narrative born from and embedded within historical and cultural contexts.

Notes

CHAPTER ONE.
ANAXIMANDER, ARCHITECTURAL
HISTORIAN OF THE COSMOS

1. Simplicius, *Physica* 24, 13; D-K 12B1.

2. Cornford 1952.

3. Heidel's study (1921, 253ff.) of the tertiary testimony on Anaximander's book led him to conclude that Anaximander was more a geographer than a philosopher, purporting to sketch the life history of the cosmos from the moment of its emergence from infinitude to the author's own time. Cherniss (1935, 18) carried this assessment further; he says that "Anaximander's purpose was to give a description of the inhabited earth, geographical, ethnological, and cultural, and the way in which it had come to be what it was." Guthrie (1962, 75) joined this chorus and remarked that this would mean that "the only part of Anaximander's doctrine on which we have any but the smallest and most doubtful bits of information, namely his cosmogony, was to him only incidental or preparatory to the main purpose of his work." And this may be so. But from this assessment of the likely content of Anaximander's book it is misleading in our effort to understand speculative and abstract thought to conclude that the book was not "philosophical." Even if we concede that the cosmological details that have been so much better preserved than other details connected with Anaximander's writings, and which almost certainly were a prelude to the practical matters to be covered in his "universal history" of the world from its creation up until the present day, it is too myopic, too narrow a conception of "philosophy"—most certainly at these earliest stages—not to see the speculative character of thought emblematic through his many interests, which he apparently regarded as inextricably interconnected. Dewey had it right when he urged us to see that philosophy stems from cultural peculiarities and embodied experiences of individuals situated in different contexts. Adopting just this approach, our narrative invites us to see philosophical thought emerging from the culture; truths do not assume a place in an eternal realm of ideas but rather take place on the very ground of human experience. The questions that Anaximander raises and addresses are not dictated ex nihilo but rather emerge in medias res.

4. Cf. Naddaf 2003 for a recent and clear statement of this position.

5. Vernant 1965/1983.

6. The scholarly literature on Anaximander has often emphasized "cosmic structure" but rarely the processes or mechanisms revealed by the structure. As we will see, the line of thought I establish reveals cosmic mechanism as a consequence of delineating cosmic structure.

7. The process is *abductive* and I will discuss this process at greater length later; cf. ch. 9, note 17.

8. Cf. Hahn 2001, ch. 2.

9. Ibid. Vitruvius VII.12.

10. For the dating of the temples, cf. Hahn 2001, ch. 2.

11. Cf. Hahn 2001, ch. 2 for the dating of Dipteros I in Samos and Artemision C in Ephesos.

12. Cf. Vitruvius 3.7; 7.16; 10.11. Cf. also Coulton 1977 (on Theodorus), 24–25, 163n51, 164n74 (on Chersiphron and Metagenes), 24, 28, 141–144, 163n68, 164n74. Cf. also Hahn 2001, ch. 2.

13. Xenophon, *Memoribilia* 4.2.8–10.

14. Cf. Hahn 2001, ch. 2, where there is a discussion of Pherecydes. But he is from Syros, a cycladaic island, and his surviving prose is centered on the theme of the marriage of "Zas and Chthonie," a work that shows him to be a *theologos* not a *phusiologos*. Let me add also that Pherecydes deserves both mention and consideration as a contributing source of inspiration for Anaximander's prose book(s), but the argument here is that the rationalizing, not mythologizing, narrative is illuminated more clearly in the light of architectural building projects and the architects' prose writings.

15. Cf. Hahn 2001, ch. 5 for a discussion of the patronage of temples in the archaic period.

16. Cf. Hahn 2003, 108–118.

17. Korres 1995, 10.

18. Heidel 1927. Cf. also *Odyssey* 3.169ff.

19. Korres 1995, 22.

20. Ibid., 20.

21. Ibid., 30.

22. Ibid., 30.

23. Cf. Hahn 2001 (esp. chapters 2 and 3), and 2003.

CHAPTER TWO.
ANAXIMANDER'S COSMIC PICTURE:
THE SIZE AND SHAPE OF THE EARTH

1. I wish to express my gratitude to Leonidas Bargelitotes, editor of the Greek journal *Skepsis*, for permission to reproduce a large portion of my essay that appeared in *Skepsis* XV/ii–iii, 2004, 372–395.

2. γυρόν.

3. στρογγύλον.

4. D-K 12A11.3, τὸ δὲ σχῆμα αὐτῆς (σχ.τῆς γῆς) ὑγρόν, στρογγύλον, χίονι λίθῳ παραπλήσιον.

5. γυρόν.

6. Kirk-Raven 1957, 134. ὑγρόν.

7. στρογγύλον.

8. χίονι.

9. κίονι.

10. λίθωι κίονι [III, 10.2].

11. στρογγύλη.

12. Cf. Robinson 1976, 114ff.; Plato, *Phaedo* 97Dff.

13. Heidel 1937, 74.

14. ξύλα στρογγύλα.

15. Theophrastus, *History of Plants* v.6.5; and again vii.4.5: στρογγυλόκαυλον.

16. Burnet 1898, 1930, 357.

17. στρογγύλη.

18. Robinson 1968, 30.

19. D-K 12B5, on the authority of Aetius III, 10,2, λίθωι κίονι τὴν γῆν προσφερῆ.

20. D-K 12A10.

21. ἔχειν δὲ τοσοῦτον βάθος ὅσον ἂν εἴη τρίτον πρὸς τὸ πλάτος.

22. Cf. Martin 1879, 66; cf. also O'Brien 1967, 425. Could the Greek mean that the height is three times the size of the diameter? Can such a case be constructed? It is doubtful but possible: τρ . . . τον can apparently also mean "thrice" (LSJ *sv tritos* III.2). The references to this meaning, however, seem rather obscure and late. But the doxographical source for Anaximander, the *Stromateis*, is also late. To defeat this reading, one would ideally come up with a more natural Greek way of saying "three times as big as the width." Smyth 354d suggests that τριττός would mean treble, but it is not cited in LSJ. τρίπλους, threefold, seems another candidate but is impossible to believe here as a copyist's mistake. On the other hand, one could have used τρίτεμοριον to make it clear that one was dealing with 1/3. Thus, the natural reading favors the proportions that the diameter is three times the height.

23. διὰ τὴν ὁμοίαν πάντων ἀπόστασιν.

24. Kirk-Raven 1957, 134; Kahn 1960, 81; Guthrie 1962, 95; McKirahan 1994, 38, and many others.

25. Lynch 2003, 372–376.

26. I would like to express my appreciation to two graduate students at the University of Cincinatti, Allison Sterrett and Jen Sacher, for a very useful research report that they prepared for me on this topic.

27. Schaus 1983, 85–89.

28. Ibid., 86.

29. Ibid., 87.

30. Cf. Diogenes Laertius II, 1–2, on the authority of Favorinus; D-K 12A1.

31. Cf. Cicero, de divinat. I, 50, 112; D-K 12A5A.

32. Yalouris 1980, 313–318, p. 314 and figure 1.

33. Ibid., 315.

34. Shaus, op. cit., 87.

35. φλοιός.

36. Hahn, 2001, ch. 4, δένδρον.

37. Hans Jucher, *Festschrift für Frank Bommer* (Mainz 1977), 195–196; LV, no. 196.

38. T. Geltzer, "Zur Darstellung von Himmel und erde auf einer Arkesilaus-Schale," *Museum Helveticum*, 36, 1979, 170–176, with plates.

39. D-K 12A11. Hippolytus, Ref. I.6,3. τῶν δὲ ἐπιπέδων ᾧ μὲν ἐπιβεβήκαμεν, ὃ δὲ ἀντίθετον ὑπάρχει.

40. Cf. Vernant 1983, 176–189, who elegantly states the matter.

41. Schaus, op. cit., 87.

42. Ibid. Schaus notes the passage in *De Hebdomadibus* 2.24.

43. The fields of archaeology and art history are especially intertwined when applied to research in the Greco-Roman world, and so I have allowed this discussion of art historical and archaeological "inference" to overlap. This close interweaving is not obvious in other fields of archaeology—for example, in new world archaeology.

44. Herodotus gives voice to just this amazement. He is clearly impressed by great engineering feats, that are therefore given a prominent role in his ethnographies. The *erga megala te kai thomasta* in the proem were traditionally taken to refer to permanent, tangible monuments (though the phrase probably encompasses deeds as well as such works). Herodotus was particularly impressed by monolithic construction, as at 2.175. A few other semirandom notices of large structures: 1.93, 1.178–187; 2.124–125, 134, 148–150, 175–176. 3.60.

45. Cf. Hahn 2001 and 2003. I have already argued extensive cases for the importance of architectural technologies to illuminate Anaximander's cosmological speculations. Here, however, I am reframing this research to produce an exemplar of a kind of argument that specifically makes the case that it is the *archaeologists' reports and artifacts* that provided the evidence.

46. Hahn 2001, ch. 2, where this discussion is presented in great detail, esp. 69–85.

47. Hahn 2003, 137ff.

48. Cf. Hahn 2001, 152–161 and 195–196.

49. This is a point completely missed by Guthrie in *A History of Greek Philosophy*, vol 1 (Cambridge: Cambridge University Press, 1971), 98–99.

50. Cf. Hahn 1996, 2001, 2003.

51. I owe this idea of the identification with a column *base* to Peter Schneider of the Technical University of Stuttgart, who has excavated with both the DAI and OAI in Didyma and elsewhere on the west coast of Turkey. I have tried to reconstruct his argument based on the evidence he drew to my attention.

52. This column base is still unpublished, but its discovery was brought to my attention by Peter Schneider.

53. Schneider 1996, 79 Abb. 3.

54. Gruben 1996. Gruben provides a general view of the development of Ionic columns (Abb. 266 on page 322f.). The comparative chart follows page 74. The same chart was published much earlier in the now classics article "Das Archaiche Didymaion," in *JdI* 78, 1963, Abb 38 after page 155. The fact is that the earliest known column bases of the seventh and first half of the sixth century are of a purely cylindrical carving.

55. Cf. Delos, Oikos of the Naxians in S. P. Courbin, *L'Oikos des Naxiens* (1980). For the column base see 44 fig. 8, plate 3 and plate 23, the base of the so-called prostoon—the entrance-hall of this building (ca. 560 BCE). The cylindrical part of the basis (*trochilos*) has a diameter of 65.2 cm and a height of 21.7 cm (plate 23), that is the proportion of exactly 3:1. See Gruben 1996, Abb. 18 after 74.

56. Gruben 1996, loc cit.

57. I owe the expressions in these last few lines to Peter Schneider.

58. Cf. Hahn 2003, 105–109.

59. ὁμοιότητα.

60. μᾶλλον μὲν γὰρ οὐθὲν ἄνω ἢ κάτω ἢ εἰς τὰ πλάγια φέρεσθαι προσήκει τὸ ἐπὶ τοῦ μέσου ἱδρυμένον καὶ ὁμοίως πρὸς τὰ ἔσχατα ἔχον.

61. Cf. Hahn 2003, 92ff, where I already anticipated this general debate and again indicated my preference for Anaximander's identification with a column *drum*.

62. Cf. Hahn 2001, ch. 2 on early prose writing.

CHAPTER THREE.
ANAXIMANDER'S COSMIC PICTURE:
THE *HOMOIOS* EARTH, '9', AND THE COSMIC WHEELS

1. As it will become clear, I am arguing that the doxographical reports are garbled. Anaximander was describing the *distances to*, not the sizes of, the heavenly wheels. Consequently, the cosmic numbers are *radii*. Cf. also the arguments in Hahn 2001, ch. 4.

2. ὁμοίως πρὸς τὰ ἔσχατα ἔχον.

3. Cf. Hippolytus 1,6.5. Although the text is damaged, the general sense is sufficiently clear: κατωτάτω δὲ τοὺς τῶν ἀπλανῶν ἀστέρων κύκλους.

4. That is, not before my recent studies; cf. Hahn 2001 and 2003.

5. McKirahan 1994, 40, and others.

6. διὰ τὴν ὁμοίαν πάντων ἀπόστασιν.

7. D-K 12A1. μέσην τε τὴν γῆν κεῖσθαι, κέντρου τάξιν ἐπέχουσαν.

8. D-K 12A2. Ἀναξίμανδρος Πραξιάδου Μιλησίου πρῶτος δὲ . . . εὗρε . . . τὴν γῆν ἐν μεσαιτάτῳ κεῖσθαι.

9. ὁμοίως δ᾽ ἔχον.

10. Plato. *Phaedo* 108E–109A; cf. also *Timaeus* 62D–63A. ἀκλινές.

11. δίνη.

12. Robinson 1971, 112. οὐρανός.

13. Furley 1987, 25.

14. ὁμοίως πρὸς τὰ ἔσχατα ἔχον.

15. He is preceded by Neuhäuser 1883, 409–421.

16. ἔκκλισις.

17. Heidel 1906, 279–292; and 1937, 7–8 and 68–69.

18. Burnet 1892/1945, 61–62.

19. Rescher 1958, 722.

20. Cf. also Heath 1913.

21. Cf. D-K 12A11: κίνησιν ἀίδιον εἶναι. Rescher and Burnet argued that the vortex arises only after the formation of the cosmos, but not before it.

22. Robinson 1971, 111–118.

23. διὰ τὴν ἰσορροπίαν καὶ ὁμοιότητα.

24. Τὴν δὲ γῆν εἶναι μετέωρον ὑπὸ μηδενὸς κρατουμένην, μένουσαν δὲ διὰ τὴν ὁμοίαν πάντων ἀπόστασιν.

25. Cf. Burnet 1898, 61 and 69ff for a defense of *dinê*.

26. Kirk-Raven-Schofield 1983, 134.

27. Kahn 1960, 53–55.

28. Guthrie 1962, 99–100.

29. Vernant 1965/1983, 180.

30. West 1971, 87.

31. Lloyd 1979, 68, but guardedly.

32. McKirahan 1994, 40.

33. Naddaf 1998.

34. Couprie 1995.

35. D-K 12A1, that the earth remains in the center.

36. D-K 12A2, that the earth remains in the center.

37. D-K 12A26, though confused.

38. Kahn 1960, 55.

39. τῶν δὲ ἐπιπέδων ᾧ μὲν ἐπιβεβήκαμεν, ὃ δὲ ἀντίθετον ὑπάρχει.

40. Vernant, 1983.

41. Cf. Hahn 2001, 169–172.

42. Hesiod, *Theogony* 726ff; cf. also my imaginative picturing of Hesiod's cosmology in Hahn 2001, 175–179.

43. D-K 21B28.

44. Vernant 1965/1983, 180ff.

45. Naddaf 2003.

46. περὶ μεγεθῶν καὶ ἀποστημάτων.

47. D-K 12A19; Simplicius discusses this matter with reference to the planets, though the evidence concerning planets in Anaximander's cosmology is exiguous.

48. Cf. Kahn 1960, 61 and 86n.1; but Dreyer 1906, 15n.1, accepts the text as it stands.

49. Tannery 1887, 91.

50. Diels 1897, 231.

51. Burnet 1892, 68.

52. Heath 1913, 38; cf. also Hahn 2003, 85–98.

53. Cf. Hahn 2003.

54. Kahn 1960, 96.

55. Burkert 1972, 41. Yet Burkert accepts the inside/outside measures, though he agrees with Kirk-Raven that there is consequently an error in computation, 309n55.

56. McKirahan 1994, 38. He also defends the view that Anaximander "assumes that the sizes and distances of the earth and heavenly bodies are related by simple proportions, with emphasis on the number three."

57. Dicks 1970, 45.

58. Cf. ch. six in this volume.

59. Kahn 1960, 90.

60. Burkert 1963.

61. West 1971, 97.

62. Schmitz 1988, 77–78: "Der umgekehrte Weg, auf dem prägende Motive Anaximanders in den Iran gelangt sein können, ist viel besser zu verfolgen."

63. Kahn 1960, 62.

64. West 1971, 86n3.

65. West 1971, 86n3.

66. Diels 1897, 231.

67. Dreyer 1906/1953, 14–15.

68. Mieli, 46.

69. Burch 1949, 154, but Burch also considers the application of the numbers in terms of radii.

70. Kahn 1960, 62.

71. Guthrie 1962, 95.

72. Zeller 6th ed. 1963, 300n2; 1st ed. 1923, 70n92.

73. Burnet 1930, 68.

74. Taylor 1928, 163.

75. Tannery 1887, *Sciene hellene*, 94.

76. Couprie 1995 and 2003.

77. Hahn 2003.

78. D-K 11A1; D.L. i. 24.

79. Cf. Heath 1913, 22.

80. Cf. Heath 1913, 311ff.

81. Heath 1913, 32.

82. Hahn 2001.

83. Hahn 2003, 105ff.

84. Hahn 2001, 76–77; 6–7.

85. Cf. Hahn 2001, ch. 2.

86. Cf. Hahn 2001, ch. 3.

87. Cf. Hahn 2001, ch. 3, where models are explored and discussed.

88. Cf. Hahn 2001, ch. 5.

89. Cf. Hahn 2003, 105ff.

90. Cf. Mertens 1991, 155–160.

91. ὁμοίως πρὸς τὰ ἔσχατα ἔχον.

92. Cf. Hahn 2001, ch.4, esp. 217–218.

93. O'Brien 1967.

94. I would like to gratefully acknowledge Sarah Taylor for her research assistance in preparing a report on "Technological Style" from which I have drawn freely in this chapter. Any errors or omissions are, of course, my responsibility alone.

95. Cf. Trigger 1989.

96. Ibid.

97. Ibid.

98. J. Sackett (1985) developed the idea of isochrestic variation, which refers to visible and invisible attributes of form, as style. Cf. also Weissner 1985.

99. Whether Dunnell would be considered a processualist is actually debatable; in many ways, his work differs from both processualists and postprocessualists.

100. Cf. Wobst 1999.

101. Cf. Flannery 1967.

102. Cf. Durkheim 1893.

103. Cf. Mauss 1936.

104. Cf. Lemonnier 1992.

105. Lemonnier 1993, 97.

106. Cf. Lemonnier 1993, 97.

107. Cf. Smith 1970, an M.I.T. historian of technology.

108. Cf. Stark 1999.

109. Cf. Smith 1970.

110. Cf. Lechtman 1977, 1984.

111. Cf. Weinberg 1965.

112. Cf. Lemonnier 1993, 2.

113. Cf. Pfaffenberger 1992.

114. Cf. Lemonnier 1993.

115. Ibid.

116. Cf. Sillar and Tite 2000.

117. Cf. Lechtman 1977.

118. Ibid., 4.

119. Cf. Weinberg 1965.

120. Cf. Lechtman 1977.

121. Lectman and Wobst, for example, do see technological style as unconscious, while Childs and Lemonnier do not.

122. Cf. Stark 1999.

123. Cf. Dobres 1999.

124. Cf. Bourdieu 1977, 79.

125. For a fascinating defense of the idea that logic is based on bodily experience, rather than vice versa, cf. Mark Johnson 1987.

126. Cf. Bourdieu 1977, 78.

127. Cf. Sutton 2001.

128. Cf. Lemonnier 1993.

129. Cf. Lemonnier 1992. Lemonnier often refers to some of the work of Leroi-Gourhan who wrote about a concept he called "technological milieu," which is very similar to representations or technological style, but may focus too much on decorative style. Unfortunately, Leroi-Gourhan's work has not been translated.

130. Cf. Lemonnier 1992.

131. Cf. Lechtman 1973, 1977; Lemonnier 1993.

132. Gruben 1996.

133. Cf. Wesenberg 1983; but see Gruben 1996, who reminds us of the many complications to the general proportions.

134. See ch. two.

135. Yalouris 1980; cf. Hahn 2003, p162n.

136. Cf. Hahn 2001, ch. 2, and 2003, 105ff.; Hahn 2004, 377–384.

137. Cf. Powell 1998, 186.

138. Cf. Hahn 2003, 121–129, for a hint of what a study might look like.

139. Finley 1965.

140. Lechtman 1977, 13.

CHAPTER FOUR.
ANAXIMANDER'S COSMIC PICTURE:
THE "BELLOWS" AND COSMIC BREATHING

1. ὥσπερ διὰ πρηστῆρος αὐλοῦ.

2. Diels 1897, 26.

3. Diels 1923, 12B4.

4. Burnet 1914, 18; 1930, 68.

5. Freeman 1957, 19.

6. Kirk-Raven, 1957, 135–136.

7. Guthrie 1962, 93.

8. Fränkel 1962, 264.

9. Brumbaugh 1964, 21.

10. Robinson 1968, 28.

11. De Vogel 1969, 8n1.

12. West 1971, 85 (cf. also 1966, *Hes.Th.* 863, 394–395).

13. Kahn 1960, 59.

14. Lloyd 1966, 313–314 and notes.

15. τὸ ἄπειρον.

16. ὕδωρ.

17. ἀήρ.

18. Daniel Graham has argued recently in *Explaining the Cosmos: The Ionian Tradition of Scientific Philosophy* (2005) that Aristotle has it wrong in attributing the doctrine of Material Monism (MM) to the Milesians—and thus also all who have followed his lead, ancient and modern; while the Ionians did claim that in the beginning there was an original stuff, that original stuff perished in the process of generating other new things: Generating Substance Theory (GST). And from this innovative interpretative starting point, Graham offers a fascinating new reading of Presocratic philosophy whole cloth. Graham offers what he regards to be arguments that are both "historically appropriate" and "philosophically coherent" to make his case, and in a separate study I will explore his claims. Right or wrong, Graham's thesis is not particularly relevant to mine. In a future study, I propose to follow through on my archaeological approach to raise a new line of inquiry as to what *also* counts as "historically appropriate" and "philosophically coherent" that Graham never considers. Anaximenes, and Anaximander, illuminate cosmic processes by appeal to material "felting" [*pilêsis*], a process that Graham acknowledges without ever investigating; can an appeal to archaeological resources on "felting" lend support to or undermine Graham's hypothesis? And *if* archaeological resources can lend clarity to traditional debates in classical scholarship, what new light does this shed on what *also* counts as evidence that is "historically appropriate" and "philosophically coherent"?

When Graham examines the doxographical reports on Anaximander's cosmology, he understands that in the beginning was the *apeiron* and, from that, by some quasibiological process, a seed is generated, and from that seed comes the opposites—hot and cold, wet and dry—and in turn the "elements" that are composed of them. Thus, hot and dry fire surrounds the cold and moist earth, like bark around a tree, and somehow gets separated off into concentric wheels of fire that we come to call the sun, moon, and stars. In Graham's take on the reports, the elements transform out of each other and perish into each other, but the *apeiron* does not seem to enter directly into these processes. And this interpretation has found support from those like Kahn, Schwabl, Vlastos, and perhaps Heidel. From this reading, Graham concludes that Anaximander does not appear to be a Material Monist, and from this reading of him Anaximenes could not have inherited MM either. It should be noted, however, that many commentators, including Kirk-Raven, Nietzsche, Schmitz, Gadamer, Freudenthal, Stokes, Hoelscher, and Couprie, interpret the only surviving fragment of Anaximander's book to read that "when things have their origin, into that they have their perishing"; it is from the *apeiron* that plurality emerges and it is back into the *apeiron* that all diverse things ultimately return. Graham's reading that the interchange is between the elements has had support, but a substantial assembly of scholars has advocated the reading that origins from and perishing into finds as its locus the *apeiron*, an interpretation consistent with Material Monism. It seems to me that this latter reading is the correct one, and this interpretation contributes to the challenges facing Graham's fascinating and thoughtful thesis about Anaximander.

19. Ἀναξιμανδρος [sc. τὰ ἄστρα εἶναι] πιλήματα ἀέρος τροχοειδῆ, πυρὸς ἔμπλεα, κατά τι μέρος ἀπὸ στομίων ἐκπνέοντα φλόγας.

20. Freeman 1959, 61, italics added for emphasis.

21. Guthrie 1962, 90.

22. Aetius III, 3, 1–2.

23. Cf. LSJ, 1463: *Placita* 3.3.1, Aristotle, *Meteorologica* 371a16; Chrysippus *Stoic.* 2, 203; Epicurus *Ep.* 2, 47U; Lucretius 6.424, 445.

24. πρηστήρ ἀνέμων.

25. West 1966, 390, *Theogony* 846.

26. Cf. Loeb edition, *Hesiod*, 1914, 1970 translation by Hugh G. Evelyn-White, 141.

27. Euripides frag. 384 ὀμμάτων ἄπο αἱμοσταγῆ πρνστῆρε ῥεύσονται κάτω.

28. Homer, *Iliad*, 17, 297.

29. Homer *Odyssey* 22.18: αὐλὸς παχύς.

30. αὐλὸς ἐκ χαλκείου.

31. Hippocrates, *de Articulatione*, 47 and 77.

32. Aristotle: for the tube of the clepsydra, cf. *Pr.* 914b14, and also GA 780b19; for the blowhole of cetacea, cf. *HA* 589b19 and *PA* 697a17; and for the funnel of the cuttlefish, cf. 524a10.

33. D-K 12A21[13], 12A22[21], 12B4. οἶον πρηστῆρος αὐλον.

34. Aetius II, 13, 7; D-K 12A18. κατά τι μέρος ἀπὸ στομίων ἐκπνέοντα φλόγας.

35. Diels 1922; but in the 5th edition, Kranz, apparently doubtful about Diels' translation, proposes his own, *Glutwindröhe*.

36. Lloyd 1966, 314n1.

37. Burnet 1930, 68.

38. Burnet 1930, n2.

39. Lloyd 1966, 314n1

40. Homer, *Iliad* 18, 468.

41. (δεύτερα δ᾽ εἰς Ἥφαιστον ἐβήσατο, παῦσε δὲ τόνγε ῥίμφα σιδηρείων τυπίδων, ἔσχοντο δ᾽ αὐτμῆς αἰθαλέοι πρηστῆρες.

42. Apollonius of Rhodes, 4, 777ff.; Cf. also Couprie (2001), who, agreeing with Fränkel, points out that Fränkel (1962, 577) denies that the meaning of πρηστῆρες here is "bellows": "Nicht 'der Blasebalg' . . . sondern der 'Glutwind' der oben aus dem Felsen aufstieg," and so, if Fränkel is right, the only evidence for Diels' translation collapses. Lloyd disagrees, 1966, 314n1.

43. Couprie 2001.

44. Ibid.

45. Diels 1879, 26: *folles* πρηστῆρας *obsolete voce nominat. Prêstêr enim ab eadem stripe pra vel par proficiscitur unde* πρῆσαι ἐύπρηστος πρημαίνειν *ducta sunt, quibus inest propria spirandi notio.*

46. πρῆσαι.

47. ἐύπρηστος.

48. πρημαίνειν.

49. Diels 1897, 26. Πρηστήρ *enim ab eadem stripe pra vel par proficiscitur unde* πρῆσαι ἐύπρηστος πρημαίνειν *ducta sunt, quibus inest propria spirandi notion.*

50. Lloyd 1966, 313n1.

51. Furley 1987, 18.

52. Cornford 1912, 7.

53. Farrington 1944, 37.

54. Gomperz 1943, J. Hist Ideas, 166n12.

55. Guthrie 1962, 67.

56. Herodotus II.25.

57. τὸ ἄπειρον.

58. Simplicius, *Physics*, 24, 17; D-K12B1. Cf. also note 18 where Daniel Graham's different translation is reported.

59. Γένεσις.

60. φθορά.

61. ἀλλήλοις.

62. Cf. Kirk-Raven 1957, 118.

63. Guthrie 1962, 80.

64. Guthrie 1962, 67.

65. Aetius I,3,4; D-K 13B2.

66. οἷον ἡ ψυχή ἡ ἡμετέρα ἀὴρ οὖσα συγκρατεῖ ἡμας, καὶ ὅλον τὸν κόσμον πνεῦμα καὶ ἀὴρ περιέχει.

67. Burnet 1914, 19.

68. Kirk-Raven, 57. This is only one of the possible construes of Aetius 1,3,4, but it captures the gist of all the other interpretations with regard to the macrocomsmic-microcosmic reasoning.

69. ψυχή.

70. Aetius IV,3.2; D-K 12A29: Ἀναξιμένης δὲ καὶ Ἀναξίμανδρος καὶ Ἀναξόγορας καὶ Ἀρξέλαος ἀερώδη τῆς ψυχῆς τὴν φύσιν εἰρήκασιν.

71. πνεῦμα.

72. Kahn 1960, 114.

73. D-K 64B5.

74. D-K 22B12, A15.

75. Cf. Wright 1995, ch. 4, on microcosm and macrocosm.

76. I would like to gratefully acknowledge the special research assistance I received from Kathleen Lynch that formed the basis of this section of the chapter, and from which I have drawn freely. Any errors, omissions, or misunderstandings are my responsibility alone.

77. See the site of Chrysokamino, Crete, among others: Betancourt 1999, especially 362–363; Rickard 1939.

78. See Betancourt 1999; compare also the topographical relationship of Laurion to Athens.

79. Rickard 1939, 94.

80. Athens, Acropolis 2134a–e, Graef and Langlotz 1925; fragments 2134b and c preserve part of the bellows.

81. Cf. wine skins, also made out of the entire goat, but rarely shown with fur.

82. See the discussion in Moore 1977, especially 329–330.

83. Paris, Cabinet des Médailles 542, ARV² 438.133, 1653, Add² 239; See Buitron-Oliver 1995, no. 1878.

84. See Gempeler 1969; see note 15 for other forge scenes in vase painting.

85. Cf. the pot bellows in figure 4.9. Nozzles are integrated into the ceramic body of the pot bellow, therefore independent nozzles must be from skin sack bellows.

86. For archaeological evidence of metalworking in Athens, see Mattusch 1977.

87. For example, Mattusch 1977, 368.

88. See Mattusch 1977, and J. Papadopoulos, Tuyères, and Kiln Firing Supports," *Hesp* 61 (1992), 207, fig. 2.

89. Heilmeyer and Zimmer 1987, catalogue II 1.

90. Reproduction in the Science Museum, South Kensington, modeled after a blacksmith's forge in Sudan, photographs in Rickard 1939, 95, figs. 2, 3, 4.

91. See Rickard 1939, 94, for ethnographic description of pot bellows.

92. See Betancourt 1999, 358–359.

93. Betancourt 1999, 358–359; deposits of fused mud on the outside of several fragments imply this arrangement.

94. Bodnar 1988, 49–51.

95. Cf. ch. two, where *Technological Style* is discussed in detail.

96. D-K 12A11.10–11.

97. Hippolytus *Ref*, 1.7.5; D-K 13A7.

98. γεγονέναι δὲ τὰ ἄστρα ἐκ γῆς διὰ τὸ τὴν ἰκμάδα ἐκ ταύτης ἀνίστασθαι, ἧς ἀραιουμένη τὸ πῦρ γίνεσθαι, ἐκ δὲ τοῦ πυρὸς μετεωριζομένου τοὺς ἀστέρας συνίστασθαι.

99. The translation is from Kirk-Raven 1957, 152, italics added for emphasis.

100. Aetius 1.31; D-G 276.

101. Hippolytus 1.74–5.

102. Aetius 2.13.14; D-K 21A38, and 21A40.

103. Hippocrates *On Breath*, 3.

104. Cf. Freeman 1959, 61.

CHAPTER FIVE.
ANAXIMANDER'S COSMIC PICTURE:
THE HEAVENLY "CIRCLE-WHEELS" AND THE *AXIS MUNDI*

1. But cf. Furley 1987, who argues Simplicius' point that Anaximander's earth stays aloft by floating on air, as Anaximenes also argued.

2. Parmenides 5–7.

3. Aetius II,20,I.

4. Ἀναξίμανδρος (sc. τὸν ἥλιον φησι) κύκλον . . . ἁμαρτείῳ τροχῷ παραπλή-σιον, τὴν ἁψῖδα ἔχοντα κοίλην, πλήρη πυρός.

5. ἁμαρτείῳ τροχῷ.

6. I have omitted from Aetius' testimony the part that the sun's circle is twenty-eight times the size of the earth. I have argued elsewhere that Aetius is mistaken, not with the assignment of numbers, but only in the implication that it is circumference or diameters that is at stake. It is not the size of the circle but the distance to the circle that Anaximander's numbers identify. Accordingly, I have argued that the distance to the outermost part of the sun's circle is twenty-eight earth-diameters. Cf. Hahn 2001, 184–187.

7. πόρους τινὰς αὐλώδεις.

8. Kirk-Raven 1957, 135.

9. Kahn 1960, 86; Mckirhan 1994, 40.

10. Kirk-Raven 1957, 135; Kahn 1960, 86.

11. Guthrie 1962, 94.

12. Hippolytus I,6.1, *kinesin aidion einai*. Cf. also Aristotle *Phys.* 250b11, and *de Caelo* B13, 295a7ff. Ps.-Plut. *Strom.* 2.

13. Simplicius, *Phys*, 24.21.

14. On the authority of Hippolytus *Ref.* I,6,4–5.

15. Aetius II,21,I.

16. Hippolytus, *Ref.* I,6,3, and Ps-Plutarch *Strom.* 2.

17. Cf. Hahn 2001, on the Ionic theory of proportions.

18. The usual view is that the module was identified with "lower column diameter" and for this technical discussion see "Proportions and Numbers." Hahn 2003.

19. On terrestrial cartography, cf. *D.L.* II, 1–2, and the *Suda* s.v., and Hahn 2001, ch. 4.

20. There is still some controversy about whether there was an agora in Anaximander's sixth-century Miletus. If not, it would be enough to direct our attention to whatever part of town was reserved for industrial production.

21. I would like to acknowledge with gratitude and appreciation Kathleen Lynch, who prepared a research report for me on ancient wheel making and from which I have drawn on freely throughout this chapter. Any mistakes or omissions are my responsibility alone.

22. Childe, 1957, 187–216, and 716–730.

23. Ibid., 207.

24. Ibid., 208.

25. Cf. Crouwel 1992, for further information and full and detailed citations to literary and archaeological sources. He examines the surviving visual evidence, summarized here, to reconstruct the structure and operation of wheeled vehicles.

26. Cf. Crouwel 1992, 34–38.

27. Fixed axle, profile view, Attic black-figured neck amphora, circa 575 BCE.

Athens, National Museum 353; Piraeus Painter, from Piraeus. *ABV* 2, *Para* 1, *Add²* 1.

28. These possibilities and their archaeological evidence are discussed by Crouwel 1992, 37–38, with notes.

29. Fixed axle, frontal view, Chalcidian black-figure neck amphora, circa 550 BC. Munich, Museum für antike Kleinkunst 594; from Vulci.

30. Fragmentary wheel nave from a fixed axle chariot. From Olympia, found in Well 20; third quarter fifth century BC. In Hayen 1980–1981, 185–189, figs. 26–28.

31. See Crouwel 1992, 37–38, n. 127–129, for details and references.

32. Crouwel 1992, 35, n. 104 for examples of depictions of six- and eight-spoked wheels.

33. See Figure 5.5.

34. Iron tires are known from Etruscan and the Halstatt-La Tène cultures; see Crouwel 1992, 37–38, n. 128 with references.

35. Crouwel 1992, 71, n. 341: "Cf. also carefully made, miniature ivory wheels with eight or seven spokes as votive offerings from the Artemision at Ephesus, Hogarth, *British Museum Excavations at Ephesus*, 1908, 168 f. nos. 42–43 with pl. XXVII: 2, 9." Eastern Greek fixed-axle chariot, terracotta revetment from Iasos, circa 550. Paris, Cabinet des Médailles. In Akrugal, *Griechische und römische Kunst in der Türkei*, München, 1987, pl. 84a (Iasos); Laviosa 1972–1973, figure 2; Åkerström 1966, pl. 16:2; L. Shear, *Sardes X.1. Architectural Terracottas*, Cambridge, MA, 1926, figure 13.

36. Fixed-axle, eight-spoked wheel, Etruscan red-figure volute krater by the Aurora Painter, from Cività Castellana (Falerii Veteres), circa 375 BC Rome, Villa Giulia 2491.

37. See Crouwel 1992, 38–39, for details.

38. Crouwel 1992, 78 n. 377–382 for citations.

39. Fixed-axle crossbar cart wheel from Olympia, Olympia Museum. First quarter or half of the fifth century BC; found in Well 17. In Hayen 1980–1981, 153–162, 183–185, figs. 8, 10–13, 20

40. Crouwel 1992, 87.

41. Fixed axle with spoked wheel, Athens, National Museum 1630. Red-figure pyxis from Eretria, circa 400 BC. In Beschi 1969–1970, figure 7.

42. See Crouwel 1992, 87, for discussion. In Tziafalias 1978, 175, figure 16.

43. Rotating axle, crossbar wheel cart. Attic black-figure amphora, fragment, from Athens, Acropolis 791. In Graef and Langlotz 1925, pl. 48, no. 791.

44. See discussion and evidence, Crouwel 1992, 87.

45. Crouwel 1992, 85, n. 432–435.

46. Cart model with disk wheels and rotating axle. Terracotta model, from Kerameikos. Athens, National Museum 4481, ninth century BCE. In Boardman 1957, pl. 3.

47. See the discussion, Crouwel 1992, 87.

48. See Crouwel 1992, pl. 18.2, Boston, Museum of Fine Arts 58.696.

49. ἁμαρτείῳ τροχῷ παραπλήσιον, τὴν ἄψιδα ἔχοντα κοίλην, πλήρη πυρός.

50. The sundial will be taken up in chapter six. But, although I am not now prepared to argue the case, had Anaximander made a moveable model, anticipating a planetarium (as Brumbaugh suggested years ago), perhaps the wheelwright might have a supplied a mechanism. We know also that Theodorus used a lathe on the column drums in the Samian Dipteros I, in the first half of the sixth century; perhaps there was a moving model of the heavens?

51. By "ceremonial columns" I mean *monolithic* ones that appear in sanctuaries, atop of which were placed sphinxes or other celebratory *agalmata*.

52. ἁμαρτείῳ τροχῷ παραπλήσιον, τὴν ἄψιδα ἔχοντα κοίλην.

53. Vitruvius 10.2.11.

54. Vitruvius 10.2.12.

55. Coulton 1977, 142.

56. D-K12A10.

57. I am drawing from the discussion of the cosmic tree directly from Schibili 1990.

58. Hesiod, *Theogony* 727ff, *Erga* 19.

59. OF, 168.29: *chthonos rizai*.

60. D-K21A47; and B28.

61. Aeschylos *Prom.* 1047.

62. Kallimachos, *Del.* 35.

63. Apollonios of Rhodes, 2.320; 605.

64. Fr.73, A11.

65. Homer 288; Herodotus 2, 138; Homer 398, 238.

66. Schibili 1990, 70–71.

67. West 1971, 55–61, esp. 59.

68. Schibili 1990, 71.

69. Plato, *Republic* X, 616Cff.

70. Schibili 1990, 72.

CHAPTER SIX.
ANAXIMANDER'S COSMIC PICTURE:
RECONSTRUCTING THE SEASONAL SUNDIAL
FOR THE ARCHAEOLOGIST'S INVESTIGATIONS

1. Diogenes Laertius, II, 1–2; D-K 12A1: "he was the first to discover a gnomon, and he set one up on the sundials in Sparta, according to Favorinus in his Universal History, to mark solstices and equinoxes; he also constructed hour-indicators. He was first to draw an outline of earth and sea, but he also constructed a celestial model."

2. *Suda s.v.*, see also Kirk-Raven (henceforth K-R), 99. "He was first to discover the equinoxes and solstices and hour-indicators . . . he introduced the gnomon and in general made an outline of geometry. He wrote "On Nature," the "Circuit of the Earth" and "On the Fixed Stars" and a "Celestial Model" and some other works.

3. Agathemerus, I,1; D-K 12A6.

4. Strabo, I, 7 (Casaubon); D-K 12A6.

5. Dicks 1970, 45; cf. also the detailed diatribe 1966, 26–40.

6. Hahn 2001, 205ff, conjectured reconstruction illustrated on 209.

7. Gibbs 1976, 6–7, and 94n12. Gibbs is referring to an archaeological report by D. S. Hunt who discovered this *analemma* in 1938 in a well on Chios. Hunt's report gives only a verbal description, which, I believe for the first time, was illustrated in my 2001 study.

8. There is plentiful evidence for identifying the shape of Anaximander's earth with a flat cylinder, expressly analogized with a column drum: D-K 12B5, 12A11.3, 12A10.

9. There is a long tradition of scholarship that emphasized the importance of "microcosmic-macrocosmic" thought to the Pythagoreans. Cf. Burkert 1972, esp. 37, 271, 294, 362, 473. But it is clear that this kind of reasoning, akin to "analogy," was made use of by Anaximander, whether or not, as one tradition would have it, he played some role in "teaching" Pythagoras. Cf also Wright 1995.

10. In preparation of column drums for installation, the *empolion* is a wood or metal dowel that is inserted in a hole in the center of a drum and allows the new drum to be aligned with the previous drum or base as it is lowered into place. Cf. Hahn 2001, pp 7–8, 154–156, 162, 196.

11. I would like to acknowledge with gratitude the preparation of many of the figures in this chapter by physicist and astronomer Richard E. Schuler.

12. Agathemerus D-K 12A6, Strabo DK 12A6, and D.L. 12A1. *Pixax* and *perimetron* and [*ges*] *periodos* almost certainly refer to the same thing: a map.

13. Cf. Clay 1992 for a consideration of a map with uninhabitable regions over and against the *oikumene*.

14. Heidel 1937, esp. 2–59. This also was my point of departure in Hahn 2001, 208ff.

15. Aristotle. *Meteorologica* II.6,

16. D-K 12A10. Ps.-Plutarch reports that the original fire surrounded the cosmos *like bark around a tree*. The cosmos was not spherical, it was cylindrical—a great tree.

17. Photo by the author, with placement of wooden gnomon added for effect. Glessmer and Albani 1999, 442: "It could have been used to handle the discrepancy between 365.25 days and a calendar year of 364 days. It allows the determination of the cardinal points and fixing a calendar whose seasons are as near as possible to the signs of sun, moon and stars."

18. For an example of a column drum from the archaic temple to Apollo at Didyma, cf. Hahn 2003, 91, fig. 2.3.

19. For a discussion of the architect's technique of *anathyrôsis*, and its implications for Anaximander's cosmic picture, cf. Hahn 2001, 149–162 and 194–195.

20. Herodotus II 109.

21. Cf. Johannes 1937.

22. According to Herodotus II, 109, the Greeks learned from the Babylonians about the celestial sphere and the gnomon and the twelve parts of the day. Cf. also K-R 1957, 102.

23. K-R, ibid.

24. Cf. Hahn 2003, 85–89; 2001, 173–174, 184–185.

25. Cf. Brumbaugh 1964, 21.

26. Cf. Mertens 1991 for the general techniques of architectural measuring with string or cord.

27. Cf. Hahn 2001, chs. 2 and 3; Hahn 2003, 73–78.

28. D-K 12A3, Aelian V.H. III, 17.

29. Cf. West 1971, 76n1.

30. Cf. Hahn 2003, 54 where Naddaf conjectures that the center of Anaximander's map was Naucratis. Naddaf stands alone among scholars in supposing that the

center of Anaximander's flat-disk earth is in Egypt, not Delphi. What he missed, however, was what the sundial would have revealed to him, namely, that the center was further south in Egypt.

31. Aristotle, *De Caelo* B 13, 295b10ff.

32. D-K 12A11.4; *Ref.* I,6, 4–5.

33. D-K 12A12; *Strom.* 2.

34. For a discussion of the lengths of a *pous* and *pexus* cf. Herodotus II 149; and Hahn 2003, 92.

35. Cf. Hahn 2003, 100–121.

36. D-K 12A1.

37. For the discussions of the numbers and proportions of the sun wheel, cf. Hahn 2003, 144–149. The debate arises from the problem of interpreting the doxography as providing either the "distances" to the wheels, in which case the numbers are radii, or rather the numbers provide the "sizes," in which case they present the circumferences of the circles. I have argued that Anaximander's first concern was emphasizing "distances"—since he was the first to argue for "depth" in astronomical space.

38. Cf. Dicks 1970, 45, 174. Dicks is the most prominent of the naysayers; see Dicks 1966 for the full argument.

39. Cf. Hahn 2003 for architectural evidence of *trisecting* angles on column-drum faces, 94–98.

40. D-K 12A1 *horoskopeia*, D.L.; *horologia*, in the *Suda*.

41. Herodotus II, 109.

42. Cf. Pauly-Wissowa s.v. Kalendar 1952, 111. But the evidence for the New Year in Sparta is less clear than these sources suggest. Samuel 1972, 108–109, seems to suggest that the year was divided into a winter semester plus a summer semester. So could that logically be: fall equinox to spring equinox plus spring equinox to fall equinox? The last month was associated with the onset of autumn (Dalios, 109). So while we can see where the deduction of autumnal equinox comes from, the conclusion about the start of the New Year in Sparta cannot be fully certain. For additional Spartan reforms, cf. 109. Cf. also Trümpy 1997, 135–140.

43. It is also possible to imagine that the sundial, even the version set up in Sparta, was much smaller. From the reports we have about the archaic architects, we know that they were capable of making miniatures. There is a famous story that Pliny relates of the architect Theodorus who supposedly made a miniature of a chariot drawn by four horses that could be hidden behind the wing of a fly. Since, as we now discuss, the architects—their techniques and their projects refined by trial and error—significantly influenced Anaximander's thoughts and writings, their model-making miniatures might also have encouraged him to produce smaller versions of the sundial in Sparta or elsewhere.

44. Cornford 1952.

45. Couprie 2003, 185–186.

46. Ibid.

47. This particular phrasing Couprie wrote to me in an exchange we had about the sundial; I wanted to be sure I understood his position.

48. Cf. Hahn 2003, 95.

49. Cf. Hahn 2003, 96.

50. Cf. Hahn 2003, 97.

51. Cf. Hahn 2003, 98, referring to Kienast's discussion at the Eupalinion.

52. I am grateful to Dirk Couprie for pointing this out to me and for supplying the illustration (fig. 6.19).

53. Never mind, for the illustration at this moment, that the earth is likened to a square shape. The point is made sufficiently clear regardless of the circular shape of the earth.

54. Diogenes Laertius, ix, 33. Cf. also the discussion in Heidel 1927, and Guthrie 1965, 422–423.

55. Kirk-Raven 1957, 413–414.

CHAPTER SEVEN.
THE PROBLEMS: ARCHAEOLOGY
AND THE ORIGINS OF PHILOSOPHY

1. Hahn 2001.

2. Hahn 2003.

3. Barnes 1979.

4. Barnes 1979, xii, italics and emphasis supplied.

5. Barnes 1982, xvi.

6. MacIntyre 1984. For example, MacIntyre makes this point clearly in the postscript to the second edition. On 265, he cites Frankena saying that we must distinguish philosophy from history, and then MacIntyre makes clear why we that cannot do this. See also 265, bottom of the page, where he argues that "evaluative and normative concepts, maxims, arguments and judgments about which the moral philosopher inquires—are nowhere to be found except as embodied in the historical lives of particular social groups and so possessing the distinctive characteristics of historical existence." See also the last full paragraph on p. 268 and read on to the top of 269, where the general point is reiterated and elaborated on. See also MacIntyre 1988, in the chapter on "The Rationality of Traditions." See the last full paragraph on p. 354 where the general thesis is propounded again.

7. Shanks and Hodder 1998, 88.

8. Hahn 1987.

9. Cf. Hahn 2001, ch. 1.

10. Cf. Lloyd 1966 as a classic example.

CHAPTER EIGHT.
WHAT IS THE ARCHAEOLOGIST'S
THEORETICAL FRAME WHEN INFERRING IDEAS

1. I would like to thank Bob Brier, Hermann Kienast, and Prudence Rice for their time in reading this chapter and making useful comments and suggestions.

2. See Wylie 2002 for an excellent overview.

3. Cf. Hahn 2001, ch. 3, 149ff., where I have discussed *anathyrôsis* in great detail, surveying the evidence from Egypt and early Greece.

4. Cf. Wright 1995, 56–74 for an interesting discussion of microcosm-macrocosm.

5. Renfrew 1994, 3; cf. also Burkert 1972 in general.

6. Shanks and Hodder 1998, 88.

7. Cf. Hodder 1999, 30–66. Ch. 3: How Do Archaeologists Reason? The chapter offers an interesting introduction to a variety of reasoning techniques that case studies show archaeologists actually use, and it also helps to demonstrate that the way archaeologists "work" and "reason" is often irrespective of their different theoretical viewpoints.

8. I would like to acknowledge the research assistance I received from one of my former students, Alex Berezow, in preparing parts of this chapter.

9. Cf. Sharer and Ashmore 1979, 478: "the cultural historical approach emphasizes synthesis based upon chronological and spatial ordering of archaeological data. . . . Once this is done, interpretation proceeds through use of either specific or general analogs as the basis for application of descriptive models, usually drawn from ethnography and history. The culmination of the interpretive process is a chronicle of events and general trends of cultural change and continuity in the prehistoric past."

10. I owe this description of culture history to Prudence Rice, who was kind enough to clarify culture history for me.

11. Hence the name "processual archaeology" that came to be applied to new archaeology once it was no longer so new.

12. Renfrew 1994, 10f., and Renfrew and Bahn 1996, 369ff.

13. Renfrew 1994, 6ff., and Renfrew and Bahn 1996, 371ff.

14. Renfrew 1996, 382, 384.

15. Ibid.

16. See Flannery and Marcus 1994.

17. Some archaeologists, however, believe that different theoretical approaches do change the way field work is undertaken. For example, in order to collect data and analyze it, the archaeologist must first decide which questions he or she wants to ask. The postprocessualist insists that there is no "objective" platform for this, but rather that the selections reflect biases and subjective issues that can never be purged.

18. Lewis-Williams 2002, 218–220.

19. Though Lewis-Williams does not refer to them using these terms.

20. Reception theory can be traced to work in literary theory of the 1960s and 1970s such as Warning 1975, Grimm 1975 and 1977, and Holub 1984.

21. There are two aspects of this reception approach. First, there are the ways a monument, event, or person was *portrayed* by the makers of scholarly or public opinion, and the ways those portrayals were *perceived* by the populace at large. The historical record shows how the monuments, events, or persons were portrayed, and so they are most readily determined. In contrast, however, the ways in which groups perceived historical events over time is much more difficult to determine.

22. The concept was introduced to archaeological disciplines by Assmann 1988, 1992, and 1997; esp. 1992, 19.

23. Assmann 1992, 30–34; 1997, 26f and 31.

24. In a different aspect of cultural memory, Pierre Nora (1989) invokes a distinction between "memory" and "histories"; memories are the events that actually happened, while histories are subjective representations of what historians believe is crucial to remember. Memory is a phenomenon that is directly related to the present; and since our perceptions of the past are influenced by changing contingencies in the present, memory itself must be seen as something fluid. Moreover, memory must be constituted by perceptions that, from the historian's point of view, are incidental or even irrelevant to the matters that are deemed essential to remember. In discussions about memory, a key point is that it is not to be understood as a merely private experience but also is part of the collective domain.

25. In discussing ethnography and ethnographic analogy, I am drawing on Carol McDavid's website www.webarchaeology.com/html/ethnogra.htm.

26. Processual archaeologists attempt to overcome this obstacle by means of "middle range theory." Middle range theory is a matter of creating "bridging arguments" (that are directly based on data to test hypotheses, etc.), but these arguments/theories are lesser or smaller than the "grand theories" of why civilizations evolve. Thus, middle range theory can be seen as attempting to bridge "data" to "grand theories. Ethnoarchaeology, the observation of contemporary groups that seem to resemble past cultures, provides data for some middle range theory. Experimentation, for example, with flintknapping, is another source of middle range information.

27. Cf. Shanks and Hodder 1998.

28. Cf. Hodder 1986, 124ff.

29. Cf.Hodder 1986, 74ff.

30. Cf. Shanks and Hodder 1998, 86f.

31. Scale and density of settlement are straightforward enough, but how the determinations of acephaly and lineage-based descent were made is not stated.

32. Hodder 1992, 45ff.

33. Hodder 1992, 171f., states this explicitly, noting that postprocessual theories of interpretation have not produced postprocessual methodologies for digging. That is, archaeological evidence is gathered in the same way under both schools of thought.

34. Shanks and Hodder 1986, 87.

35. Shanks and Hodder 1986, 86.

36. Shanks and Hodder 1986, 86.

CHAPTER NINE.
THE INTERPRETATIVE MEANING OF AN OBJECT:
GROUNDING HISTORICAL NARRATIVES
IN LIVED EXPERIENCE

1. I want to record my appreciation for and gratitude to Dr. John Kaag for his preparation of a research report that became the foundation of this chapter, and from which I have drawn freely. We worked together formulating the substance of this study, and I have drawn extensively from his report. Any mistakes, omissions, and misinterpretations are mine alone.

2. This desire to employ a particular historical approach to truth and to mediate between two philosophical camps is reflected by Thomas Kuhn in *The Essential Tension* (1977) when he writes: "Increasingly, I suspect that anyone who believes that history may have deep philosophical import will have to learn to bridge the longstanding divide between the Continental and English-language philosophical traditions" (xv).

3. Wylie 2002, 6, 14, 36, 93–97,

4. Bernstein 1983, 124.

5. Gadamer makes this point clear in his description of historical consciousness. See "The Problem of Historical Consciousness" in Paul Rabinow, *Interpretative Social Science: A Reader*. Berkeley: University of California Press, 1979, 151–152.

6. Wylie 2002, 172–173.

7. Bernstein 1983, 139.

8. Translation provided by Martin Heidegger in Heidegger 1998, 361.

9. Gadamer 1975, 130.

10. Gadamer 1975, 131.

11. Cf. ch. eight, note 26.

12. Dewey 1981/1925, 361.

13. Ibid., 11.

14. Dewey 1986/1938, 107.

15. Bertrand Russell voices this critique of Dewey at multiple points, most notably perhaps in his *History of Western Philosophy*. See Russell 1945, 824–825.

16. Ibid., 111.

17. Abduction, in the writings of the later Peirce, is the moment of logical insight that is the sole inferential source of ampliative knowledge. This is so because abduction is the cognitive means of assimilating generality into our conscious experience beyond the level of understanding we gain simply through perceptual judgment. Abduction is a flash of insight that arranges the generality we encounter in perceptual experience into a configuration (a hypothesis) that, when it is true, expands our understanding of generals and their relationship with facts. For the sake of explication, I will situate abduction in relation to the other forms of inference, articulate how abduction relates to generality, and explain Peirce's distinction between abduction and perceptual judgment. I am grateful to Jason Rickman for guiding me through this issue, and have drawn freely from a research report he prepared for me.

Logic is, for Peirce, one of the three normative sciences, and its role is to relate the phenomenon of argumentative knowing to the end of logical goodness (i.e., to truth) [*Essential Peirce*, Vol. II, edited by the Peirce Edition Project (Bloomington: Indiana University Press, 1998), 204. Henceforth, I will refer to this text as *EP* II and then give the page(s) I am citing]. There are three forms of inference: abduction, deduction, and induction. Of these three, deduction is the only form of inference whose conclusions are necessary when it has been validly executed. Deduction is mathematical reasoning, which "starts with a hypothesis, the truth or falsity of which has nothing to do with the reasoning," and yields a conclusion that is no less ideal than its hypothesis (*EP* II:205). Induction is concerned with "the experimental testing of a theory," in order to "[measure] the degree of concordance of that theory with fact" (*EP* II:205). Neither deduction nor induction introduces new ideas to thought, for that is abduction's role: "abduction consists in studying facts and devising a theory to explain them" (*EP* II:205). Abduction suggests a highly fallible account of why things are as they are ("maybe *x* is the reason I find *y*"), while deduction articulates the necessary consequences of the abduction without making any claims about whether the hypothesis under consideration really is true (i.e., determines what the world should look like if *x* should turn out to be the case), and induction shows the degree to which the hypothesis really explains *y* (i.e., examines the facts of the world and determines whether these facts are signs that suggest *x* is something actually operative in the world that causes *y*) by showing the extent to which the experimental (experiential) facts of the world coincide with what we should expect the world to look like (i.e., the features of the world deduced from explanation *x*).

Abduction is the sole logical function that gives us new knowledge of the world, for deduction is strictly explicative (because it just tells us the necessary consequences of something), and induction is strictly evaluative. ("Observe that neither Deduction nor Induction contributes the smallest positive item to the final conclusion of inquiry. They render the indefinite definite: Deduction explicates; Induction evaluates: that is all. . . . [E]very plank of [science's] advance is first laid by Retroduction [another word Peirce uses for abduction] alone, that is to say, by the spontaneous conjectures of instinctive reason; and neither Deduction nor Induction contributes a single new concept to the structure" (*EP* II:443). Abduction is a weak form of inference in that hypothesis forming is highly fallible (hence the need for deduction and induction), yet it is because of its relationship with generality that abduction can be ampliative. To understand why generality's relationship with abduction is central to the ampliative nature of abduction, we must consider Peirce's scholastic realism.

Peirce is an Aristotelian in the sense that he is a moderate realist: our concepts are real (i.e., not useful fictions we have created), but are not existent in the world in the way that a particular horse or plant is. Peirce was an avid student of medieval logic, and he explicitly identifies his position as scholastic realism in several places. ("A great variety of thinkers call themselves Aristotelians, even the Hegelians, on the strength of special agreement. No modern philosophy [in the sense of a philosophy drawing upon or building upon the early moderns] or very little has any real right to the title. I should call myself an Aristotelian of the scholastic wing, approaching Scotism, but going much further in the direction of scholastic realism. The doctrine of Aristotle is distinguished from substantially all modern philosophy by its recognition of at least two grades of being. That is, besides *actual reactive existence*, Aristotle recognizes a germinal being, an *esse in potential* or I like to call it an *esse in futuro*. In places Aristotle has glimpse of a distinction between *energeia* and *entelecheia*" (EP II:180). He thinks that the world is full of generality, and abduction is a sort of insight into generality.

Generality is part of the category he calls Thirdness; another way he glosses Thirdness is reasonability, lawfulness. "I do not define the reasonable as that which accords with men's natural way of thinking, when corrected by careful consideration; although it is a *fact* that men's natural ways of thinking are more or less reasonable. I had best explain myself by degrees. By reasonableness, I mean, in the first place, such unity as reason apprehends—say, generality. By generality, I suppose you mean that different events resemble one another.' Humph! Not quite; let me distinguish. The green shade over my lamp, the foliage I see through the window, the emerald on my companion's finger, have a resemblance. It consists in an impression I get on comparing those and other things, and exists by virtue of their being as they are. But if a man's whole life is animated by a desire to become rich, there is a general character in all his actions, which is not caused by, but is formative of, his behavior. 'Do you mean then that there is a purpose in nature?' I am not insisting that it is a purpose, but it is the law that shapes the event, not a chance resemblance between the events that constitutes the law. 'But are you so ignorant as not to know that generality only belongs to the figments of the mind?' That would seem to be my condition. If you will have it that generality takes its origin in the mind alone, that is beside the question. But if things can only be *understood* as generalized, generalized they really and truly are; for no idea can be attached to a reality essentially incognizable" (EP II:72n). It is only because of a certain orderliness, the presence of certain (actively) ordering relationships between qualia and force, that the universe is intelligible to us (to the degree that it is intelligible). In the absence of order, we should be completely unable to adapt to the world, since, in a world of radical Heraclitean flux, the past would be of no use to us in anticipating the future (or understanding the present). Rather, Peirce thinks that generality is both real and active throughout the world in the form of lawfulness. (It should be noted that Peirce distinguishes between real and existent: "whatever exists, *ex-ists*, that is, really acts upon other existents, so obtains a self-identity, and is definitely individual," whereas "that is *real* which has such and such characters, whether anybody thinks it to have those characters or not" EP II:342.) Generals are predicative (consider the old definition of generals: *generale est quod natum aptum est dici de multis*) (EP II:183). Elsewhere, he says "'Real' is a word invented in the thirteenth century to signify having Properites, i.e. characters sufficing to identify their subject, and possessing these whether they be anywise attributed to it by any single man or group of men, or not. Thus, the substance of a dream is not Real, since it was such as it was, merely in that a dreamer so dreamed it; but the fact of the dream is Real, if it was dreamed; since if so, its date, the dame of the dreamer, etc., make up a set of circumstances sufficient to distinguish it from all other events; and these belong to it, i.e., would be true if predicated of it, whether A, B, or C" (*EP* II:435), and as such, mediate among individuals of a class.

Now, Peirce contends that our experience is shot through with generality, and that it is only on the basis of a more or less instinctual attunement with generality that has evolutionarily developed that we are able to be animals who abduct. Peirce thinks that scholastic realism is a simpler explanation than nominalism because "it is the simpler hypothesis in the sense of the more facile and natural, the one that instinct suggests, that must be preferred[,] for the reason that unless man have a natural bent in accordance with nature's he has no chance of understanding nature, at all" (*EP* II:444). All of our concepts arise from perceptual judgments, but these judgments contain general elements that we abduce (and, to some extent, test inductively), from which arise such predictive power as we have, whether it is about gravitation, what a dog is, or what a person's character is. For the abduction of a particular person's generality (qua the particular individual he or she is), see Peirce's account of his knowledge of Theodore Roosevelt at *EP* II:222. For the abduction of the general term "dog," see Perice's analysis at *EP* II:222–223. Peirce is offering an account of our acquisition of generals that bears some resemblance to Aristotle's treatment of this issue at *Posterior Analytics* II:19. That is, our knowledge of generals comes to us through sense perception, and generals are real (i.e., not helpful fictions, as the nominalist would content). A salient differences between Aristotle and Peirce are that Peirce thinks that Roosevelt has generality not only as a human but also as Roosevelt (i.e., that I can abductively know Teddy Roosevelt as the particular man he is, not just to the extent that I can subsume him within the class of homo sapiens). Another difference is that Peirce's account of this faculty is evolutionary: Just as the instincts of other animals are adaptations that these species have developed in tandem with the development of the world, so too is the abductive habit something that we do instinctively. It is only because we as a species have developed this ability that any particular person can become extremely skilled at abduction; if it were a matter of unguided trial and error, hardly anyone in his or her lifetime would become an even moderately successful abducer. Moreover, abduction and perceptual judgment lie along the same continuum: "I think it of great importance to recognize . . . that the abductive faculty, whereby we divine the secrets of nature, is, as we may say, a shading off, a gradation of that which in its highest perfection we call perception" (*EP* II:224). In perceptual judgment, we encounter generality, and it is our ability to accommodate ourselves to generality that has enabled our success as a species. The difference between perceptual judgment and abduction is a difference of degree: My making of perceptive judgments "is the result of a process . . . not controllable and therefore not fully conscious" (*EP* II:227), whereas I can have some (albeit very weak) modicum of control over abduction. "The abductive suggestion comes to us like a flash. It is an act of *insight*, although of extremely fallible insight. It is true that the different elements of the hypothesis were in our minds before; but it is the idea of putting together what we had never before dreamed of putting together which flashes the new suggestion before our contemplation" (*EP* II:227).

18. Dewey 1986/1934, xvi.

19. Ibid., 271.

20. Ibid., 101–104.

21. Ibid., 276.

22. Ibid., 292.

23. Quine 1980/1953, 44.

24. Ibid.

25. Ibid.

26. Murphy 1990, 84.

27. Quine 1969, 28–29.

28. Ibid., 38.

29. Ibid.

30. Davidson 1999, 73.

31. Davidson 1984a, 200.

32. Davidson 1984b, 326.

33. Putnam 1981, esp. 60–64.

34. Cf. Hahn 1988, 7–9.

35. Putnam 1987, 1.

36. Ibid., 33.

37. Ibid., 35.

38. Ibid., 17.

39. I owe the phrasing in this last paragraph to Mark Johnson, who was kind enough to help me clarify Putnam's position.

40. Searle 1995, 149–150.

41. Ibid., 55.

42. Ibid., 183.

43. Ibid., 183.

44. Ibid., 188–189.

45. Ibid., 195.

46. Here I am following the extremely clear presentation of Searle's arguments in Wisnewski 2005, 74–81.

47. Searle 1995, 80.

48. Ibid., 186.

49. Wisnewski 2005, 80.

50. Ibid., 79.

51. Kuhn 1962.

52. Goodman 1978.

CHAPTER TEN.
THE EMBODIED GROUND OF ABSTRACT
AND SPECULATIVE THOUGHT

1. I would like to gratefully acknowledge John Kaag with whom I worked in formulating this chapter. He produced a research report from which I have drawn freely. Any mistakes or omissions and misinterpretations are mine alone.

2. Dewey, "The Types of Philosophic Thought" 1981/1925, 24.

3. Dewey 1981/1925, 43.

4. Dewey, "The Sources of a Science Education, Individualism Old and New" 1981/1925, 163.

5. James 1900, 4.

6. Ibid., 230.

7. Johnson, "Body in Mind" 1987.

8. Ibid., 25–28; See also Lakoff 1987, 440–444. Gibbs and Colston (1995) concluded that these embodied patterns are established early in child development and are stable across cultures. Sinha (2000) elaborates on the formation of particular image schemas, suggesting that their formation is due in large part to the sociocultural forces at play.

9. Ibid., 48. See also the discussion of prototype effects in Lakoff and Johnson 1999.

10. Lakoff and Johnson 1999, 45.

11. Ibid., 58.

12. Davidson 2006, 223.

13. Lakoff and Johnson 1999, 58.

14. Kauffman 2000, 135.

15. Damasio 2001, 59.

16. The term "embodied mind" took hold in the biological sciences in the work of Francisco Varela 1979.

CHAPTER TWELVE.
THE APPLICATION OF ARCHAEOLOGY
TO ANCIENT PHILOSOPHY:
METAPHYSICAL FOUNDATIONS
AND HISTORICAL NARRATIVES

1. I would like to acknowledge thoughtful help from many colleagues but most especially Mark Johnson, Randy Auxier, Doug Anderson, Pat Manfredi, and Jason Rickman, who graciously read some of this chapter and spoke with me at length about issues arising from it. Any errors in content or presentation are mine alone.

2. Cf. Putnam 1981, ch. 4, who uses this technical parlance of the "similitude theory of reference" and the "correspondence theory of truth" in explicating Kant.

3. Hahn 1988, 7–9.

4. Here I am drawing on Perec 1978.

5. Cf. Hahn 2001, ch. one.

6. I already discussed "reception history" and "memory culture" in ch. two in connection with Lewis-Williams' discussion about the Franco-Cantabrian cave paintings.

7. White 2004.

8. Dewey 1981, 232–233.

9. But it also recognizes that there are standards of judgment to be found in the context in which narratives are offered.

10. It is useful to reflect on the historical curiosity whereby Anaximander survived, however meagerly, while other thinkers/writers did not. The philosopher Chrysippos was reputed to have written 500 words a day, producing more than 700 works (Diogenes Laertius, *Lives* VII, 780–782). Yet only a few fragments remain. In all of Aristotle's corpus, so many texts reach us more than two millennia later but not a single dialogue survives of the many he was reported to have written. And yet every title attributed to Plato survives! How many of the most brilliant (and unappreciated?) technical writings disappeared because of an inability to find a wide enough audience to understand and preserve them, or were lost by some other unforeseeable contingency? How many thinkers and authors who contributed to the origins and development of early Greek philosophy have been lost to us? The selection of whose writing survived, was transmitted, and placed together with others of supposedly like-minded spirits must be a most fascinating compilation of stories—even if it could ever be told; the tales they would surely tell would be as fascinating as they were serendipitous.

11. Cf. White 2004, 94.

12. Mead 1929, 235–242.

13. Cf. Hodder 1992, 176–179.

Bibliography

Aaboe, A. "Scientific Astronomy in Antiquity." In D. G. Kendal, F. R. Hodson et al. (eds.), *The Place of Astronomy in the Ancient World*, 21–45. Oxford: Oxford University Press, 1974.

———. "Observation and Theory in Babylonian Astronomy." *Centaurus* 24 (1980): 13–35.

Aeschylus, *Elektra*. Oxford: Oxford University Press, 1974.

Åkerström, A. *Die architektonischen Terrakotten Kleinasiens.* Lund: CWK Gleerup, 1966.

Akurgal, Ekrem. "The Early Period and the Golden Age in Ionia." *American Journal of Archaeology* 66 (1962): 369–379.

Andronikos, Manolis. "Samos: the Heraeum." Translated by F. Maxwell Brownjohn In Evi Melas (ed.), *Temples and Sanctuaries of Ancient Greece*, 179–190. London: Thames and Hudson, 1973.

Aristotle. *De Caelo.* Oxford: Clarendon Press, 1978.

———. *Metaphysics.* Oxford: Clarendon Press, 1975.

Asmis, Elizabeth. "What Is Anaximander's *Apeiron?*" *Journal of the History of Philosophy* 19 (1981): 279–297.

Assmann, Jan. Kollektives Gedächtnis und kulturelle Identität. In J. Assmann and T. Hölscher (eds.), *Kultur und Gedächtnis*, 9–19. Frankfurt: Suhrkamp, 1988.

———. *Das kulturelle und Gedächtnis. Schrift, Erinnerung und politische Identität in frühen Hochkulturen.* Munich: Beck, 1992.

———. *Moses the Egyptian. The Memory of Egypt in Western Monotheism.* Cambridge: Harvard University Press, 1997.

Austin, M., and P. Vidal-Naquet. *Economic and Social History of Ancient Greece.* 2nd ed. Berkeley: University of California Press, 1977.

Baldry, H. C. "Embryological Analogies in Presocratic Cosmogony." *Classical Quarterly* 26 (1932): 27–34.

Bammer, Anton. *Das Heiligtum der Artemis von Ephesos*. Graz: Akademische Druck-u. Verlagsanstalt, 1984.

Barnes, Jonathan. *The Presocratic Philosophers*. 2 vols. London: Routledge and Kegan Paul, 1979. 2nd ed. 1982.

Barnes, Jonathan, ed. *Early Greek Philosophy*. Harmondsworth, England: Penguin, 1987.

Bernstein, Richard. *Beyond Objectivism and Relativism—Science, Hermeneutics and Praxis*. Philadelphia: University of Pennsylvania Press, 1983.

Beschi, L. "Rilievi votive attici ricomposti." *Annuario della Scuola Archeologica di Atene e della Missioni Italiane in Oriente* 47–48 (1969–1970).

Betancourt, Phillip P., et al. "Research and Excavation at Chrysokamino, Crete, 1995–1998." *Hesperia* 68 (1999): 343–370.

Bickerman, E. J. *Chronology of the Ancient World*. Ithaca: Cornell University Press, 1968.

Binford, L. "Archaeology as Anthropology." *American Antiquity* 11 (1962): 198–200.

——— . *An Archaeological Perspective*. New York: Seminar Press, 1972.

——— . *In Pursuit of the Past: Decoding the Archaeological Record*. London: Thames and Hudson, 1983a.

——— . *Working at Archaeology*. New York: Academic Press, 1983b.

Boardman, John. "Early Euboean Pottery and History." *Annual of the British School at Athens* 52 (1957).

——— . *Greek Sculpture: The Archaic Period*. New York: Oxford University Press, 1978.

——— . *The Greeks Overseas: Their Early Colonies and Trade*. Revised ed. London: Thames and Hudson, 1980.

Bodnar, I. M. "Anaximander's Rings." *Classical Quarterly* 38 (1988): 49–51.

——— . "Anaximander on the Stability of the Earth." *Phronesis* 37 (1992): 336–342.

Bourdieu, P. *Outline of a Theory of Practice*. Translated by R. Nice. New York: Cambridge University Press, 1977.

Brumbaugh, Robert S. *The Philosophers of Greece*. New York: Crowell, 1964.

——— . *Ancient Greek Gadgets and Machines*. New York: Crowell, 1966.

Buitron-Oliver, Diana. *Douris: A Master Painter of Athenian Red Figure Vases*. Mainz am Rhein: Philipp von Zabern, 1995.

Burkert, Walter. "Iranisches bei Anaximander." *Rheinisches Museum fur Philologie*, 106 (1963): 97–134.

——— . *Lore and Science in Ancient Pythagoreanism*. Translated by E. L. Minar. Cambridge: Harvard University Press, 1972.

——— . *Homo Necans: The Anthropology of Ancient Greek Sacrificial Ritual and Myth*. Translated by P. Bing. Berkeley: University of California Press, 1983.

———. "Itinerant Diviners and Magicians: A Neglected Element in Cultural Contacts." In Robin Hagg (ed.), *The Greek Renaissance in the 8th Century B.C.*, 115–119. Stockholm: Svenska Institut, 1983.

———. *Greek Religion*. Translated by J. Raffan. Cambridge: Harvard University Press, 1985.

———. *The Orientalizing Revolution: Near Eastern Influence on Greek Culture in Early Archaic Age*. Translated by M. Pinder and W. Burkert. Cambridge: Harvard University Press, 1992.

Burnet, John. *Early Greek Philosophy* (1st ed. 1892). London: Macmillan, 1930 (4th ed.).

———. *Greek Philosophy. Part 1, Thales to Plato*. London: Macmillan, 1914.

Cherniss, Harold F. *Aristotle's Criticism of Pre-Socratic Philosophy*. Baltimore: Johns Hopkins Press, 1935.

———. "The Characteristics and Effects of Presocratic Philosophy." *Journal of the History of Ideas* 12 (1951): 319–345.

Childe, Gordon. "Wheeled Vehicles." In Charles Singer, E. J. Holmyard, and A. R. Hall (eds)., *A History of Science and Technology*. New York and London: Oxford University Press, 1954, 716–729.

Clarke, D. "Archaeology: The Loss of Innocence." *Antiquity* 47 (1973): 6–18.

———. *Analytical Archaeology*, 2nd ed. London: Methuen, 1978.

Clarke, M. L. "The Architects of Greece and Rome." *Architectural History* 6 (1963): 9–22.

Classen, Carl Joachim. "Anaximandros." *Pauly-Wissowa, Realencyclopaedia der classischen Alterumswissenschaft*, suppl. 12, 1970, cols. 30–69. Reprinted in *Ansaetze: Beitrage zum Verstandnis der fruhgriechischen Philosophie*. Wurzburg/Amsterdam, 1986, 47–92.

Clay, Diskin. "The World of Hesiod." In Apostolos N. Athanassakis (ed.), *The Ramus Essays on Hesiod II, Ramus* 21.2 (1992): 131–155.

Collingwood, R. G. *The Idea of History*. Oxford: Clarendon Press, 1946.

Conche, Marcel. *Anaximandre: Fragments et Temoignages*. Paris: Presses Universitaires de France, 1991.

Cornford, F. M. "Mystery Religions and Pre-Socratic Philosophy." In *Cambridge Ancient History*, volume IV. Cambridge: Cambridge University Press, 1939.

———. "Was the Ionian Philosophy Scientific?" *Journal of Hellenic Studies* 62 (1942): 1–7.

———. *Principium Sapientiae: The Origins of Greek Philosophical Thought*. Cambridge: Cambridge University Press, 1952.

Coulton, J. J. "Towards Understanding Greek Temple Design: General Considerations." *Annual of the British School at Athens* 70 (1975): 59–99.

———. *Ancient Greek Architects at Work*. Ithaca: Cornell University Press, 1977.

Couprie, Dirk. "The Visualization of Anaximander's Astronomy." *Apeiron* 28.3 (1995): 159–181.

———. "Anaximander's Numbers, or the Discovery of Space." Pp. 23–48 in *Essays in Ancient Greek Philosophy, VI: Before Plato*. Albany: State University of New York Press, 1998.

———. "*Presteros Aulos* Revisited." *Apeiron* 34.3 (2001): 193–202.

Crouwell, J. H. *Chariots and other Wheeled Vehicles in Iron Age Greece*. Holland: Allard Pierson Museum, 1992.

Damasio, Antonio. "Some Notes on Brain, Imagination and Creativity." In K. Pfenninger and V. Shubik (eds.), *The Origins of Creativity*. Oxford: Oxford University Press, 2001.

Davidson, Donald. "The Method of Truth in Metaphysics," in *Inquiries into Truth and Interpretation*, Vol. 1. Oxford: Oxford University Press, 1984a, 199–212.

———. "The Structure and Content of Truth," in *Inquiries into Truth and Interpretation*, Vol. 2. Oxford: Oxford University Press, 1984b, 57–74.

———. "Reply to Peter Pagin." In Urszula Zeglen (ed.), *Donald Davidson: Truth, Meaning and Knowledge*. New York: Routledge, 1999.

———. *Essential Davidson*. Eds. Donald Davidson, Ernie Lepore, Kirk Ludwig. Oxford: Oxford University Press, 2006.

De Vogel, C. J. *Greek Philosophy: A Collection of Texts*. Vol. I. Leiden: E. J. Brill, 1969.

Dewey, John. *Experience and Nature*. 1925. Vol. 1 of *John Dewey: The Later Works, 1925–1953*. Edited by Jo Ann Boydston. Carbondale: Southern Illinois University Press, 1981.

———. *Art as Experience*. 1934. Vol. 10 of *John Dewey: The Later Works, 1925–1953*. Edited by Jo Ann Boydston. Carbondale: Southern Illinois University Press, 1986.

———. *Logic: The Theory of Inquiry*. 1938. Vol. 12 of *John Dewey: The Later Works, 1925–1953*. Edited by Jo Ann Boydson. Carbondale: Southern Illinois University Press, 1986.

———. "Individualism Old and New." In Vol. 5 of *John Dewey: The Later Works, 1925–1952*. Edited by Jo Ann Boydston. Carbondale: Southern Illinois University Press, 1984, 41–144.

———. "The Sources of a Science Education" In Vol. 5 of *John Dewey: The Later Works, 1925–1952*. Edited by Jo Ann Boydston. Carbondale: Southern Illinois University Press, 1984. 1–40.

———. "Syllabus: Types of Philosophic Thought." In Vol. 13 of *John Dewey: The Later Works, 1925–1953*. Edited by Jo Ann Boydston. Carbondale: Southern Illinois University Press, 1981, 348–397.

Dicks, D. R. "Thales." *Classical Quarterly* 9 (1959): 294–309.

———. "Solstices, Equinoxes, and the Presocratics." *Journal of Hellenic Studies* 86 (1966): 26–40.

————. *Early Greek Astronomy to Aristotle*. Ithaca: Cornell University Press, 1970.

Diels, Hermann. *Doxagraphi Graeci*. 1897. Reprint, Berlin: de Gruyter, 1965.

————. "Über Anaximanders Kosmos." *Archiv für Geschichte der Philosophie* 10 (1897): 228–237.

————. "Anaximandros von Milet." *Neue Jahrbucher für das klassische Altertum, Geschichte und Deutsche Literatur* 51 (1923).

Diels, Hermann, and Walter Kranz. *Die Fragmente der Vorsokratiker* [D-K], 6th ed. Berlin: Weidmann, 1951–1952.

Dinsmoor, W. B. *The Architecture of Ancient Greece*. 3rd ed. 1902. Reprint, New York: Norton, 1975.

Diodorus Siculus. *Historical Library*. Translated by C. H. Oldfather. Cambridge: Harvard University Press, 1962.

Diogenes Laertius. *Lives of the Philosophers*. 2 vols. Translated by R. D. Hicks. Cambridge: Harvard University Press, 1966.

Dobres, Marcia-Anne. "Technology's Links and Chaines: The Processual Unfolding of Technique and Technician." In M. Dobres and G. Hoffman (eds.), *Social Dynamics of Technology: Practice, Politics, and World View*. Washington, DC: Smithsonian Institution Press, 1999.

Dreyer, J. L. E. *A History of Astronomy from Thales to Kepler*. 1906. Reprint, New York: Dover, 1953.

Duchesne-Guillemin, J. "D'Anaximandre a Empedocle: Contacts Greco-Iraniens." *Atti del convegno sul tema: la Persia a il mondo Greco-Romana* (11–14 April 1965). Rome, 1966.

Durkheim, Emile. *The Division of Labor in Society*. Cambridge: Cambridge University Press, 1893.

Farrington, B. *Greek Science*. London: Penguin Books, 1944 (vol. 1); 1949 (new edition); 1949 (vol. 2); first publication as one vol. 1953, rev. ed. 1961.

Ferguson, John. "Two Notes on the Preplatonics." *Phronesis* 9 (1964): 98–106.

————. "The Opposites." *Apeiron* 3 (1969): 1–17.

————. "Dinos." *Phronesis* 16 (1971): 97–115.

Finley, M. I. "Technological Innovation and Economic Progress in the Ancient World." *The Economic History Review*, series 2, 18 (1965): 29–45.

Flannery, Kent. "Culture History vs. Culture Process: A Debate in American Archaeology." *Scientific American* 217 (1967): 119–122.

Flannery, Kent, and J. Marcus. "Ancient Zapotec Ritual and Religion: An Application of the Direct Historical Approach." In C. Renfrew and E. B. W. Zubov (eds.), *The Ancient Mind: Elements of Cognitive Archaeology*, 55–74. Cambridge: Cambridge University Press, 1994.

Frankel, Hermann. *Early Greek Poetry and Philosophy*. Trans. Moses Hadas and James Willis. New York and London: Harcourt Brace Jovanovich. 1962/1975.

Frankfort, Henri. *Before Philosophy: The Intellectual Adventure of Ancient Man.* Harmondsworth, England: Penguin, 1949.

Freeman, Kathleen. *Companion to the Presocratic Philosophers.* Cambridge: Harvard University Press, 1959.

Furley, David. *The Greek Cosmologists.* Vol 1. Cambridge: Cambridge University Press, 1987.

————. *Cosmic Problems.* Cambridge: Cambridge University Press, 1989.

Furley, David, and R. E. Allen, eds. *Studies in Presocratic Philosophy. Vol. I: The Beginnings of Philosophy.* New York: Humanities Press, 1970.

Gadamer, Hans Georg. "The Problem of Historical Consciousness." In Paul Rabinow (ed.), *Interpretative Social Science: A Reader.* Berkeley: University of California Press, 1979. 103–160.

————. *Truth and Method.* Translated by Garrett Barden. New York: Seabury Press, 1975.

Geltzer, T. "Zur Darstellung von Himmel und Erde auf einer Arkesilaus-Schale." *Museum Helveticum* 36 (1979): 170–176.

Gempeler, R. "Die Schmeide des Hephäst—Eine Satyrspieleszene des Harrow-Malers." *Antike Kunst* 12 (1969): 16–21.

Gernet, L. *The Anthropology of Ancient Greece.* 1917. Translated by J. Hamilton and B. Nagy. Baltimore: Johns Hopkins University Press, 1981.

Gibbs, Raymond, and Herbert Colston. "The Cognitive Psychological Realities of Image Schemas and Their Transformations." *Cognitive Linguistics* 6.4 (1995): 347.

Gibbs, Sharon L. *Greek and Roman Sundials.* New Haven: Yale University Press, 1976.

Gigon, Olof. *Der Ursprung der griechishen Philosophie: Von Hesiod bis Parmenides.* 2nd ed. Basel: Schwabe, 1968.

Glessmer, Ewe, and Matthias Albani, "An Astronomical Measuring Instrument from Qumran." In Donald W. Parry and Euguene Ulrich (eds.), *The Provo International Conference on the Dead Sea Scrolls, Technological Innovations, New Texts, and Reformatted Issues.* Boston: Brill, 1999.

Goodman, Nelson. *The Ways of Worldmaking.* New York: Hackett, 1978.

Graef, B., and E. Langlotz. *Die antiken Vasen von der Akropolis zu Athen, unter Mitwirkung von Paul Hartwig, Paul Wolters, Robert Zahn; veröffentlicht von Botho Graef.* Berlin: Walter de Gruyter, 1925.

Graham, Daniel W. *Explaining the Cosmos: The Ionian Tradition of Scientific Philosophy.* Princeton: Princeton University Press, 2005.

Grimm, Gunter. Einführung in die Rezeptionsforschung. In G. Grimm, *Literatur und Leser, Theorien und Modelle zur Rezeption literarischer Werke,* 11–84. Stuttgart: Reclam, 1975.

————. *Rezeptionsgeschichte. Grundlegung einer Theorie.* Munich: Wilhelm Fink, 1977.

Gruben, Gottfried. "Das archaische Didymaion." *Jahrbuch des deutschen archaologischen Institut* 78 (1963): 78–177.

———. "Greichische Un-Ordnung." In *Säule und Gebälk, Diskussionen zur Archäologisichen Bauforschung*, Band 6 (1996), 61–77. Mainz am Rheim: Philipp von Zabern, 1996.

———. *Die Tempel der Griechen*. Munich: Hirmer Verlag, 1976.

Guthrie, W. K. C. *Aristotle: On the Heavens*. London: Heinemann, 1939; repr. 1953.

———. *A History of Greek Philosophy*. Vol. 1. Cambridge: Cambridge University Press, 1962.

Hahn, Robert. "A Note on Plato's Divided Line." *Journal of the History of Philosophy* 21.2 (1983): 235–237.

———. "What Did Thales Want To Be When He Grew-Up? or, Re-Appraising the Roles of Engineering and Technology on the Origins of Greek Philosophy/Science." In Brian Hendley (ed.), *Plato, Time, and Education: Essays in Honor of Robert S. Brumbaugh*, 107–130. Albany: State University of New York Press, 1987.

———. "Kant's Newtonian Revolution in Philosophy." *Journal of the History of Philosophy*, Monograph Series. Carbondale: Southern Illinois University Press, 1988.

———. "Anaximander and the Architects," in *Proceedings of the Society for Ancient Greek Philosophy*, presented to the annual meeting of the American Philological Association, New Orleans, December 1992.

———. "Technology and Anaximander's Cosmical Imagination: A Case Study for the Influence of Monumental Architecture on the Origins of Western Philosophy/Science." In Joseph C. Pitt (ed.), *New Directions in the Philosophy of Technology*, 93–136. Netherlands: Kluwer, 1995.

———. *Anaximander and the Architects: The Contributions of Egyptian and Greek Architectural Technologies to the Origins of Greek Philosophy*. Albany: State University of New York Press, 2001.

———. "Imagining Philosophical Rationality: A Case Study of Archaeology's Contribution to Early Greek Philosophy." *Skepsis* 15.2–3 (2004): 372–395.

———. "Heidegger, Anaximander, and the Greek Temple," Electronic text, Supplementary Theorie der Architectur. http://www.tucottbus.de/BTU/Fak2/ TheoArch/Wolke/eng/Subjects/071/Hahn/hahn.htm.

Hahn, Robert, Dirk Couprie, and Gerard Naddaf. *Anaximander in Context: New Studies in the Origins of Greek Philosophy*. Albany: State University of New York Press, 2003.

Hannah, Robert. *Greek and Roman Calendars: Constructions of Time in the Ancient World*. London: Duckworth, 2005.

Hayen, H. "Zwei in Holz erhalten geblieben Reste von Wagenräder aus Olympia." *Die Kunde* n.s. 31–32 (1980–1981), figs 26–28.

Heath, T. L. *The Thirteen Books of Euclid's Elements*. 3 vols. 2nd ed. Cambridge: Cambridge University Press, 1926.

———. *Aristarchus of Samos: The Ancient Copernicus*. 1913. Reprint, Oxford: Clarendon Press 1959.

———. *A History of Greek Mathematics*. 1921. Reprint, Oxford: Oxford University Press, 1965.

Hegel, G. W. F. *Lectures on the History of Philosophy*. 2nd ed. 3 vols. 1892. Translated by E. S. Haldane. Reprint, London: Routledge and Kegan Paul, 1963.

Heidegger, Martin. *Early Greek Thinking*. Translated by D. F. Krell and F. A. Capuzzi. New York: Harper and Row, 1975.

———. *Pathmarks*. Cambridge: Cambridge University Press, 1998.

Heidel, W. A. "The DINE in Anaximenes and Anaximander." *Classical Philosophy* 1 (1906): 279–282.

———. "Anaximander's Book and the Earliest Known Geographical Treatise." *Proceedings of the American Academy* (1921): 7.

———. *The Frame of the Ancient Greek Maps. With a Discussion of the Discovery of the Sphericity of the Earth*. American Geographical Society Research Series, 20. New York: American Geographical Society, 1937.

Heilmeyer, W. D. and G. Zimmer. "Die Bronzegiesserei unter der Werkstatt des Phidias in Olympia." *Archäologischer Anzeiger* 102 (1987): 239–299.

Herodotus. *Histories*. Oxford: Oxford University Press, 1984.

Hesiod. *Theogony*. Oxford: Oxford University Press, 1977.

———. *Theogony and Other Works*. Translated by Glenn W. Most. Cambridge: Harvard University Press, 2007.

Hesse, Mary. *The Structure of Scientific Inference*. London: Macmillan, 1974.

Hodder, I. *Reading the Past*. Cambridge: Cambridge University Press, 1986.

———. *Theory and Practice in Archaeology*. London: Routledge, 1992.

———. *The Archaeological Process*. Oxford: Blackwell, 1999.

Holloway, R. R. "Architect and Engineer in Archaic Greece." *Harvard Studies in Classical Philosophy* 73 (1969): 281–290.

Holscher, Uvo. "Anaximander and the Beginnings of Greek Philosophy." *Hermes* 81 (1953): 255–277 and 385–417.

Holub, Robert C. *Reception Theory: A Critical Introduction*. London and New York: Methuen, 1984.

Hunt, D. W. S. "An Archaeological Survey of the Classical Antiquities on the Island of Chios Carried Out between the Months of March and July 1938." *The Annals of the British School at Athens* 41 (1940–1945): 41–42.

James, William. *The Principles of Psychology*. New York: Holt, 1892.

Johannes, H. "Die Saulenbasen vom Heratempel des Rhoikos." *Mitteilungen des deutschen archaologischen Instituts, Athenische Abteilung* 62 (1937): 13–37.

Johnson, M. *Archaeological Theory*. Oxford: Blackwell, 1999.

Johnson, Mark. *The Body in the Mind*. Chicago: University of Chicago Press, 1987.

Kahn, Charles H. *Anaximander and the Origins of Greek Cosmology*. New York: Columbia University Press, 1960.

———. "On Early Greek Astronomy." *Journal of Hellenic Studies* 90 (1970): 99–116.

Kauffman, Stuart. *Investigations*. Oxford: Oxford University Press, 2000.

Kienast, H. J. "Der Sog. Tempel D im Heraion von Samos." *Mitteilungen des Deutschen Archaologischen Instituts, Athenische Abteilung* 100 (1985): 105–127.

———. "Der Tunnel des Eupalinos auf Samos." *Mannheimer Forum* (1986/1987): 179–241.

———. "Fundamentieren in schwierigem Gelande; Fallstudien aus dem Heraion von Samos." In *Bautechnik der Antike: Diskussion zur Archaologischen Bauforschung*, 123–127. Band 5, Mainz am Rhein: Verlag Phillipp von Zabern, 1991.

———. "Die Basis der Geneleos-Gruppe." *Mitteilungen des Deutschen Archaologischen Instituts, Athenische Abteilung* 107 (1992): 29–42.

———. *Die Wasserleitung des Eupalinos auf Samos*. *Samos XIX*. Mainz am Rhein: Philipp von Zabern, 1995.

Kirk, G. S. "Some Problems in Anaximander," *Classical Quarterly* 5 (1955): 21–38.

Kirk, G. S., J. E. Raven, and Malcolm Schofield. *The Presocratic Philosophers*. 2nd ed. Cambridge: Cambridge University Press, 1983.

Korres, Manolis. *From Pentelicon to the Parthenon: The ancient quarries and the story of a half-worked column capital of the first marble Parthenon*. Athens: "Melissa," 1995.

Kranz, Walther. "Vorsokratisches I." *Hermes* 69 (1934): 114–119.

———. "Vorsokratisches II." *Hermes* 69 (1934): 226–228.

———. "Kosmos als philosophischer Begriff fruhgriechischer Zeit." *Philologus* 93 (1938): 430–448.

Kuhn, Thomas. *The Structure of Scientific Revolutions*. Chicago: University of Chicago Press, 1962.

———. *The Essential Tension*. Chicago: Chicago University Press, 1977.

Lakoff, George. *Women, Fire and Dangerous Things: What Categories Reveal about the Mind*. Chicago: University of Chicago Press, 1987.

Lakoff, George, and Mark Johnson. *Philosophy in the Flesh: The Embodied Mind and its Challenge to Western Thought*. New York: HarperCollins, 1999.

Laviosa, C. "Un rilievo arcaico di Iasos e il problema del fregio nei templi ionici." *Annuario della Scuola Archeologica di Atene e della Missioni Italiane in Oriente* 50–51 (1972–73): 397–418.

Lawrence, A. W. *Greek Architecture*. 5th ed. revised by R. A. Tomlinson. New York: Penguin, 1996.

Leach, E. R. "Primitive Time-Reckoning." In Charles Singer, E. J. Holmyard, and A. R. Hall (eds.), *A History of Technology*, Vol 1, 110–127. New York: Oxford University Press, 1954.

Lechtman, Heather. "The Gilding of Metal in Pre-Columbian Peru." In W. J. Young (ed.), *Application of Science in Examination of Works of Art*. Boston: 1973.

―――. "Style in Technology-Some Early Thoughts." In H. Lechtman and R. S. Merrill (eds.), *Material Culture: Styles, Organization and Dynamics of Technology*. St. Paul: West, 1977.

―――. "Andean Value Systems and the Development of Prehistoric Metallurgy." *Technology and Culture* 25.1 (1984): 1–36.

Le Goff, Jacques. *History and Memory*. New York: Columbia University Press. 1992.

Lemonnier, Pierre. *Elements for an Anthropology of Technology*. Anthropological Papers, no. 88, Museum of Anthropology, University of Michigan, Ann Arbor, 1992.

―――. "Introduction." In *Technological Choices: Transformation in Material Culture Since the Neolithic*. London: Routledge, 1993.

Lewis-Williams, J. David. *A Cosmos in Stone: Interpreting Religion and Society Through Rock Art*. Walnut Creek, CA: Alta Mira Press, 2002.

Littman, R. J. *The Greek Experiment: Imperialism and Social Conflict 800–400 B.C.* London: Thames and Hudson, 1974.

Lloyd, G. E. R. "Hot and Cold, Dry and Wet in Early Greek Thought." *Journal of Hellenic Studies* 84 (1964): 92–106.

―――. *Polarity and Analogy: Two Types of Argumentation in Early Greek Thought*. Cambridge: Cambridge University Press, 1966.

―――. *Magic, Reason, and Experience: Studies in the Origins and Development of Greek Science*. Cambridge: Cambridge University Press, 1979.

―――. *Demystifying Mentalities*. Cambridge: Cambridge University Press, 1990.

Lynch, Kathleen, M. "When Is a Column Not a Column? Columns in Attic Vase-Painting." *Acta of the 2003 International Archaeological Congress*, ed. Amy Brauer and Carol Mattusch, pp. 372–376.

MacIntyre, Alasdair. *After Virtue*. 2nd edition. Notre Dame, IN: University of Notre Dame Press, 1984.

―――. *Whose Justice? Which Rationality?* Notre Dame, IN: University of Notre Dame Press, 1988.

Martin, Roland. *Manuel D'Architecture Grecque*. Paris: J. Picard, 1965.

Mattusch, C. "Bronze- and Ironworking in the Area of the Athenian Agora." *Hesperia* 46 (1977): 340–379.

Mauss, Marcel. "Les Techniques du Corps." In *Sociology and Psychology: Essays of Marcel Mauss*, translated by B. Brewster. London: Routledge and Kegan Paul, 1979.

McKirahan, Richard D. Jr. *Philosophy Before Socrates*. Indianapolis: Hackett, 1994.

Mead, George H. "The Nature of the Past." In J. Coss (ed.), *Essays in Honor of John Dewey*. New York: Holt, 1929, 235–242.

Mertens, Dieter. "Schnurkonstruktionen." In *Bautechnik der Antike*, vol 5, 155–160. Mainz am Rhein: Philipp von Zabern, 1991.

Mitchell, B. M. "Cyrene and Persia." *Journal of Hellenic Studies* 86 (1966): 99–113.

Moore, M. B. "The Gigantomachy of the Siphnian Treasury: Reconstruction of Three Lacunae." *Études delphiques* (BCH Suppl. IV) (1977): 305–335.

Morrison, J. S. "Pythagoras of Samos." *Classical Quarterly* 50 (1956): 135–156.

Mugler, Charles. *Deux Themes de la Cosmologie Greque: Devenir Cyclique et Pluralite des Mondes*. Paris: Librairie C. Klincksierck, 1953.

Murphy, John. *Pragmatism: From Peirce to Davidson*. Boulder: Westview Press, 1990.

Naddaf, Gerard. "Anaximander's Measurements Revisited." *Essays in Ancient Greek Philosophy, VI: Before Plato*. Albany: State University of New York Press, 1998, 5–22.

———. "On the Origins of Anaximander's Cosmological Model," *Journal of the History of Ideas* 59(1) (1998): 1–28.

Naddaf, Gerard, Robert Hahn, and Dirk Couprie. *Anaximander in Context: New Studies in the Origins of Greek Philosophy*. Albany: State University of New York Press, 2003.

Neugebauer, O. "Ancient Mathematics and Astronomy." In C. Singer, E. J. Holmyard, and A. R. Hall (eds.), *A History of Technology*, 785–804. New York and London: Oxford University Press, 1954.

———. *The Exact Sciences in Antiquity*. 2nd ed. Providence: Brown University Press, 1957.

———. *A History of Ancient Mathematical Astronomy*. 3 vols. Berlin: Springer, 1975.

Neuhäuser, I. *Dissertatio de Anaximandri Milesius sive vetustissima quaedem rerum universitatis conceptio restituta*. Bonnae: Max Cohen et Filius, 1883.

Nora, Pierre. "Between Memory and History: Les Lieux de Mémoire," *Representations* 26 (1989): 7–25.

O'Brien, D. "Anaximander's Measurements." *The Classical Quarterly* 17 (1967): 423–432.

Orlandos, A. *Les Materiaux de Construction: Et la Technique Architecturale des Anciens Grecs*. 2 vols. Translated by V. Hadjimichali. Paris: Editions E. De Boccard, 1966 and 1968.

Pauly, A. F., G. Wissova, and W. Kroll. *Realencylopädie der classicschen Altertumswissenschaft*. Stuttgart: 1894, and following years.

Pausanius. *Guide to Greece*. 2 vols. Translated by Peter Levi. New York: Penguin, 1971.

Perec, George. *Life: A User's Manual*. London: Harvill, 1978.

Pfaffenberger, Bryan. "Social Anthropology of Technology." *Annual Review of Anthropology* 21 (1992): 491–516.

Pindar. *Pythian Odes*. Oxford: Clarendon Press, 1970.

Pindar. *Isthmian Odes*. Oxford: Clarendon Press, 1970.

Plato. *Philebus*. Oxford: Clarendon Press, 1976.

——— . *Republic*, vol 4. Oxford: Clarendon Press 1980.

Pliny the Elder. *Natural History*. Translated by H. Rackham. Cambridge: Harvard University Press, 1938.

Plutarch. *Lives*. Translated by B. Perrin. Cambridge: Harvard University Press, 1968.

Putnam, Hilary. *Reason, Truth, and History*. Cambridge: Cambridge University Press, 1981.

——— . *The Many Faces of Realism*. Peru, IL: Open Court, 1987.

Quine, W. V. O. *From a Logical Point of View*. 1953. Cambridge: Harvard University Press, 1980.

——— . *Ontological Relativity and Other Essays*. New York: Columbia University Press, 1969.

——— . *The Pursuit of Truth*. 1990. Cambridge: Harvard University Press, 1992.

Rankin, H. D. "*Homoios* in a Fragment of Thales." *Glotta* 39 (1961): 73–76.

Renfrew, C. "Towards a Cognitive Archaeology." In C. Renfrew and E. B. W. Zubov (eds.), *The Ancient Mind: Elements of Cognitive Archaeology*, 3–12. Cambridge: Cambridge University Press, 1994.

Renfrew, C., and P. Bahn. *Archaeology: Theories, Methods, and Practice*. New York: Thames and Hudson, 1996.

Renfrew, C., and E. B. W. Zubov, eds. *The Ancient Mind: Elements of Cognitive Archaeology*. Cambridge: Cambridge University Press, 1994.

Rescher, Nicholas. "Cosmic Evolution in Anaximander?" *Studium Generale* 11: 718–731.

Rickard, T. A. "The Primitive Smelting of Iron." *American Journal of Archaeology* 43 (1939): 85–101.

Robertson, D. S. *Greek and Roman Architecture*. 2nd ed. Cambridge: Cambridge University Press, 1943.

——— . *A Handbook of Greek and Roman Architecture*. 2nd ed. Cambridge: Cambridge University Press, 1969.

Robin, Leon. *Greek Thought and the Origins of the Scientific Spirit*. New York: Russell & Russell, 1967.

Robinson, John Mansley. *An Introduction to Early Greek Philosophy*. Boston: Houghton Mifflin, 1968.

——— . "Anaximander and the Problem of the Earth's Immobility." In J. Anton and G. Kustas (eds.), *Essays in Ancient Philosophy*, vol. 1, 111–118. Albany: State University of New York Press, 1971.

Rochberg, Francesca. "Astronomy and Calendars in Ancient Mesopotamia." In J. M. Sasson (ed.), *Civilizations of the Ancient Near East*. Peabody, MA: Hendrickson, 1995. Vol III, 1925–1940.

Russell, Bertand. *A History of Western Philosophy*. New York: Simon and Schuster, 1945.

Sackett, J. "Style and Ethnicity in the Kalahari: A Reply to Weissner." *American Antiquity* 50 (1985): 154–159.

Saltzer, W. "Vom chaos zu Ordnung: Die Kosmologie der Vorsokratiker." In U. Schultz (ed.), *Die wissenschaftliche Eroberung des Kosmos*, 61–70. Munich: Beck, 1990.

Sambursky, S. *The Physical World of the Greeks*. 3 vols. Translated by Merton Dagut. Princeton: Princeton University Press, 1956.

Samuel, Alan E. *Greek and Roman Chronology: Calendars and Years in Classical Antiquity*, Munich: Beck'sche Verlagbuchhandlung, 1972.

Sarton, George. *A History of Science*. Vol 1. London: Harvard University Press, 1952.

Schaus, G. P. "Two Notes on Lakonian Vases." *American Journal of Archaeology* 87 (1983): 85–89.

Schibli, Hermann. *Pherecydes of Syros*. Oxford: Clarendon Press, 1990.

Schmitz, H. *Anaximander und die Anfaege der griechischen Philosophie*. Bonn: Bouvier Verlag, 1988.

Schneider, Peter. "Untersuchuungen an der Terrassenmauer im Apollon-Bezirk von Didyma." *Istanbuler Mitteilungen* 34 (1984): 326–343.

———. "Neue Funde vom archaishen Apollontempel in Didyma." *Säule und Gebälk', Diskussionen zur Archäologisichen Bauforschung*, Band 6 (1996), 78–83.

———. "Zum Alten Sekos von Didyma." *Istanbuler Mitteilungen* 46 (1996): 147–152.

Searle, John R. *The Construction of Scoial Reality*. New York: Free Press, 1995.

Shanks, M., and I. Hodder. "Processual, Post-Processual and Interpretive Archaeologies." In D. S. Whitley (ed.), *Reader in Archaeological Theory*, 69–98. New York: Routledge, 1998.

Shanks, M., and C. Tilley. *Re-Constructing Archaeology: Theory and Practice*. Cambridge: Cambridge University Press, 1987.

———. *Social Theory and Archaeology*. Albuquerque: University of New Mexico Press, 1988.

Sharer, Robert J., and Wendy Ashmore. *Fundamentals of Archaeology*. Menlo Park, CA: Benjamin Cummings, 1979.

Shear, L. *Sardis X.1. Architectural Terracottas*. Cambridge, England: American Society for the Excavation of Sardis, 1926.

Shipley, Graham. *A History of Samos 800–188 B.C.* Oxford: Clarendon Press, 1987.

Siegel, R. E. "Hestia: On the Relation Between Early Greek Scientific Thought and Mysticism—Is Hestia, the Central Fire, an Abstract Astronomical Concept?" *Janus* 49 (1960): 1–20.

Sillar, B., and M. S. Tite. "The Challenge of 'Technological Choices' for Materials Science Approaches in Archaeology." *Archaeometry* 42.1 (2000): 2–20.

Singer, Charles et al., eds. *A History of Technology.* Vols. I and II. Oxford: Clarendon Press, 1957.

Sinha, Christopher, and K. Jensen de Lopez. "Language, Cultural Context, and the Embodiment of Spatial Cognitions." *Cognitive Linguistics* 11.2 (2000): 14–41.

Smith, Cyril Stanley. "Art, Technology and Science: Notes of their Historical Interaction." *Technology and Culture* 11 (1970): 493–549.

Snell, Bruno. *The Discovery of the Mind.* Translated by T. G. Rosenmeyer. Cambridge: Cambridge University Press, 1953.

———. *Archaic Greece: Age of Experiment.* Berkeley and Los Angeles: University of California Press, 1980.

Solmsen, Friedrich. "Anaximander's Infinite: Traces and Influences." *Archiv für Geschichte der Philosophie* 44 (1962): 109–131.

Stark, M. T. "Social Dimensions of Technological Choice in Kalinga Ceramic Tradition." In Elizabeth S. Chilton (ed.), *Material Meanings: Critical Approaches to the Interpretation of Material Culture.* Salt Lake City: University of Utah Press, 1999.

Stokes, Michael C. "Hesiodic and Milesian Cosmogonies." *Phronesis* 7 (1962): 1–37; 8 (1963): 1–34.

Sutton, David. *Remembrance of Repasts: An Anthropology of Food and Memory.* Oxford: Berg, 2001.

Szabo, A. "Anaximandros und der Gnomon." *Acta Antiqua* 25 (1977): 500–529.

Tannery, Paul. *Pour l'histoire de la science hellene: De Thales a Empedocle.* 2nd ed. Paris: Gauthier-Villars, 1887/1930.

Thomson, J. Oliver. *History of Ancient Geography.* Cambridge: Cambridge University Press, 1948.

Thomson, George. *Studies in Ancient Greek Society: The First Philosophers.* London: Lawrence and Wishart, 1955.

Thrower, Norman Joseph William. *Maps and Man: An Examination of Cartography in Relation to Culture and Civilization.* Englewood, NJ: Prentice-Hall, 1972.

Thucydides. *Histories.* Oxford: Oxford University Press, 1978.

Tomlinson, R. A. *Greek Sanctuaries.* London: Paul Elek, 1976.

———. *Greek Architecture.* London: Bristol Classical Press, 1989.

Trigger, Bruce. *A History of Archaeological Thought.* New York: Cambridge University Press, 1989.

Trümpy, Catherine. *Untersuchungen zu den altgriechischen Monatsnamen und Monatsfolgen.* Heidelberg: Universitat Verlag C. Winter, 1997.

Tziafalias, A. "Excavations at Hagios Georgios near Larissa." *Athens Annals of Archaeology* 11 (1978).

Varela, Francisco. *The Principles of Biological Autonomy.* New York: Elsevier, 1979.

Vernant, Jean-Pierre. *The Origins of Greek Thought.* Translation of *Les Origines de la pensee greque*, Paris, 1962. Ithaca: Cornell University Press, 1982.

———. *Myth and Thought in Ancient Greece*. Translation of *Mythe et pensee chez les grecs*, 2nd ed., Paris 1965. London: Routledge & Kegan Paul, 1983.

Vidal-Naquet, P. "La raison greque et la cite." *Raison Presente* 2 (1967): 51–61.

Vitruvius. *The Ten Books on Architecture*. Translated by M. H. Morgan. New York: Dover, 1968.

Vlastos, Gregory. "Equality and Justice in Early Greek Cosmologies." *Classical Philology* 42 (1947): 156–178.

———. "Theology and Philosophy in Early Greek Thought." *Philosophical Quarterly* 2 (1952): 97–123.

———. "Isonomia." *American Journal of Philology* 74.4 (1953): 337–366.

von Fritz, Kurt. *Grundprobleme der Geschichte der antiken Wissenschaft*. Munich: Walter De Gruyter, 1971.

———. *Science Awakening*. 3rd ed. Translated by A. Dresden. New York: Oxford University Press, 1961.

Walter, Hans. *Das Griechische Heiligtum: Heraion von Samos*. Munich: R. Piper, 1965; 2nd enlarged edition, Stuttgart: Urachhaus, 1990.

———. *Das Heraion von Samos: Ursprung und Wandel eines griechischen Heiligtums*. Munich: R. Piper, 1976.

Warning, Rainer (ed.). *Rezeptionsästhetik*. Munich: Wilhelm Fink. 1975.

Weinberg, S. S. "Ceramics and the Supernatural: Cult and Burial Evidence in the Aegean World." In F. Matson (ed.), *Ceramics and Man*. Chicago: Aldine, 1965.

Weissner, P. "Style or Isochrestic Variation? A Reply to Sackett." *American Antiquity* 50 (1985): 160–166.

Wesenberg, B. Review of Petronotis' *Zum Problem der Bauzeichnungen bei den Griechen*. *Gnomon* 48 (1976): 797–802.

———. *Beitrage zur Reconstruktion griechischer Architektur nach literarischen Quellen*, Mitteilungen des Deutschen Archaologischen Instituts, Athenische Abteilung 9, Beiheft. Berlin: Gebr. Mann Verlag, 1983.

West, M. L. *Hesiod's Theogony*. Oxford: Clarendon Press, 1966.

———. *Early Greek Philosophy and the Orient*. Oxford: Clarendon Press, 1971.

White, K. D. *Greek and Roman Technology*. London: Thames and Hudson, 1984.

Wisnewksi, J. Jeremy. "Rules and Realism: Remarks on the Poverty of Brute Facts." *Sorites* (16 December 2005): 74–81.

Wobst, H. M. "Style in Archaeology or Archaeological Style." In Elizabeth S. Chilton (ed.), *Material Meanings: Critical Approaches to the Interpretation of Material Culture*. Salt Lake City: University of Utah Press, 1999.

Wright, M. R. *Cosmology in Antiquity*. London: Routledge, 1995.

Wylie, Allison. *Thinking Through Things: Essays in the Philosophy of Archeology*. Berkeley: University of California Press, 2002.

Yalouris, N. "Astral Representations in the Archaic and Classical Periods and their Connection to Literary Sources." *American Journal of Archaeology* 84 (1980): 313–318.

Xenophon. *Memoribilia.* Oxford: Oxford University Press. 1981.

Zaehner, R. C. *The Dawn and Twilight of Zoroastrianism.* London: Weidenfeld and Nicolson, 1961.

Zeller, Edward. *Die Philosophie der Griechen.* 1923. Edited and enlarged by W. Nestle. Hildesheim: G. Olms, 1963.

Index

Made in the USA
Lexington, KY
22 April 2018